DEADLY COMPANIONS

DOROTHY H. CRAWFORD is Professor of Medical Microbiology and Assistant Principal for the Public Understanding of Medicine at the University of Edinburgh. Among numerous scientific publications, she is the author of *The Invisible Enemy: A Natural History of Viruses* (OUP, 2000). She is a Fellow of the Royal Society of Edinburgh, a Fellow of the Academy of Medical Sciences, and was awarded an OBE in 2005 for services to medicine and higher education.

DEADLY COMPANIONS

How Microbes Shaped Our History

D O R O T H Y H. C R A W F O R D

OXFORD

UNIVERSITY PRESS

OXFORD
UNIVERSITY PRESS

Great Clarendon Street, Oxford OX2 6DP

Oxford University Press is a department of the University of Oxford.
It furthers the University's objective of excellence in research, scholarship,
and education by publishing worldwide in

Oxford New York

Auckland CapeTown Dar es Salaam Hong Kong Karachi
Kuala Lumpur Madrid Melbourne Mexico City Nairobi
New Delhi Shanghai Taipei Toronto

With offices in

Argentina Austria Brazil Chile Czech Republic France Greece
Guatemala Hungary Italy Japan Poland Portugal Singapore
South Korea Switzerland Thailand Turkey Ukraine Vietnam

Oxford is a registered trade mark of Oxford University Press
in the UK and in certain other countries

Published in the United States
by Oxford University Press Inc., New York

© Dorothy H. Crawford 2007

The moral rights of the author have been asserted
Database right Oxford University Press (maker)

First published 2007

First published in paperback 2009

All rights reserved. No part of this publication may be reproduced,
stored in a retrieval system, or transmitted, in any form or by any means,
without the prior permission in writing of Oxford University Press,
or as expressly permitted by law, or under terms agreed with the appropriate
reprographics rights organization. Enquiries concerning reproduction
outside the scope of the above should be sent to the Rights Department,
Oxford University Press, at the address above

You must not circulate this book in any other binding or cover
and you must impose the same condition on any acquirer

British Library Cataloguing in Publication Data

Data available

Library of Congress Cataloging in Publication Data

Data available

Typeset by SPI Publisher Services, Pondicherry, India
Printed in Great Britain
on acid-free paper by
Clays Ltd, St Ives plc

978-0-19-956144-5

6

Contents

CONTENTS

Figures and Tables

FIGURES

TABLES

Preface

Microbes first appeared on planet Earth around 4 billion years ago and have coexisted with us ever since we evolved from our ape-like ancestors. By colonizing our bodies these tiny creatures profoundly influenced our evolution, and by causing epidemics that killed significant numbers of our predecessors they helped to shape our history. Through most of this coexistence our ancestors had no idea what caused these 'visitations' and were powerless to stop them. Indeed the first microbe was only discovered some 130 years ago and since then we have tried many ingenious ways to stop them from invading our bodies and causing disease. But despite some remarkable successes, microbes are still responsible for 14 million deaths a year. In fact new microbes are now emerging with increasing frequency, while old adversaries like tuberculosis and malaria have resurged with renewed vigour.

This book explores the links between the emergence of microbes and the cultural evolution of the human race. It combines a historical account of major epidemics with an up-to-date understanding of the culprit microbes. Their impact is discussed in the context of contemporary social and cultural events in order to show why they emerged at particular stages in our history and how they caused such devastation.

We begin with SARS, the first pandemic of the twenty-first century. Then we travel back in time to the origin of microbes and see how they evolved to infect and spread between us with such apparent ease. From there we follow the interlinked history of man and microbes from the 'plagues' and 'pestilence' of ancient times to the modern era, identifying key factors in human cultural changes from hunter-gatherer to farmer to city-dweller which made us vulnerable to microbe attack.

The final chapters show how modern discoveries and inventions have impacted on today's global burden of infectious diseases and ask how, in an ever more crowded world, we can overcome the threat of emerging microbes. Will pathogenic microbes be 'conquered' by a 'fight to the death' policy? Or is it time to take a more microbe-centric view of the problem? Our continued disruption of their environment will inevitably lead to more conflict with more microbes, but now that we appreciate the extent of the problem surely we can find a way of living in harmony with our microscopic cohabitants of the planet.

Throughout this book the term 'microbe' is used for any organism that is microscopic, be it a bacterium, virus, or protozoan (Figure 0.1). Fungi are also included since although their vegetative growth is often visible to the naked eye, the spores that spread them from one host to another are microscopic. None of these tiny life forms have brains, so despite the fact that they often appear ingenious and manipulative, they have no facilities to think or plan. The human characteristics often attributed to them actually come about by their ability to adapt rapidly to changing situations. Then the natural process of 'survival of the fittest' ensures that the best adapted prosper, so that in the end they really

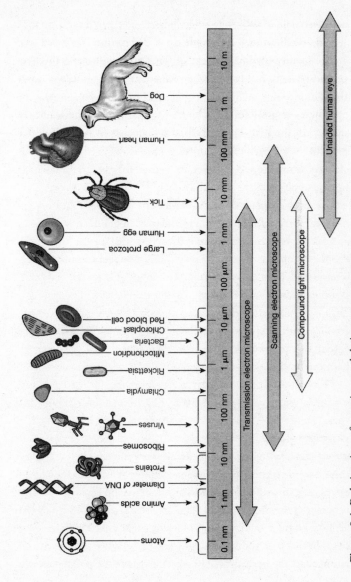

Figure 0.1 Relative sizes of organisms and their component parts

Source: J. G. Black, *Microbiology, Principles and Exploration*, 5th edn © 2002, Fig. 3.2; reprinted with permission of John Wiley & Sons, Inc.

do seem to lie in wait for a suitable host and then 'jump', 'attack', 'invade' and 'target'. Although these descriptions seem apt and are frequently used in the text of this book to illustrate the lives of microbes, in reality microbes are not capable of malice afore-thought.

Wherever possible the scientific terms used in the book are defined in the text, but there is also a glossary at the end which provides additional information.

Acknowledgements

This book could not have been written without help and support from many people to whom I am extremely grateful. In particular I thank my editor, Latha Menon, for her help and encouragement, and the following colleagues for providing expert information and advice: Professor Sebastian Amyes (antibiotic resistance), Dr Tim Brooks (plague), Dr Helen Bynum (historical events), Professor Richard Carter (malaria), Dr Gareth W Griffith (potato blight), Professor Shiro Kato (smallpox in Japanese cultural history), Dr Francisca Mutapi (schistosomiasis), Dr G Balakrish Nair (cholera), Professor Tony Nash (flu), Dr Richard Shattock (potato blight), Professor Geoff Smith (smallpox), Dr John Stewart (bacteria), Dr Sue Welburn (trypanosomiasis), Professor Mark Woolhouse (epidemiology). In addition I am grateful to the following for reading and commenting on the manuscript: Danny Alexander, William Alexander, Martin Allday, Roheena Anand, Jeanne Bell, Cathy Boyd, Rod Dalitz, Ann Guthrie, Ingo Johannessen, Karen McAulay, J. Alero Thomas.

I am also indebted to Dr Ingo Johannessen for virological research, John and Ann Ward for organizing a visit to Eyam, Elaine Edgar for literature research, Sir Anthony Epstein for facilitating research on the history of smallpox, and Dr Tasnim Azim for hosting my visit to the International Centre for Diarrhoeal Disease Research, Bangladesh.

Finally, I thank the University of Edinburgh for granting me a sabbatical year to research and write the manuscript.

INTRODUCTION

When SARS first hit an unsuspecting world in 2003 the press had no need to dramatize or embellish. The true story certainly rivalled any modern thriller with a mysterious killer virus on the loose in Southern China, innocently carried in a human incubator to its international launch pad in Hong Kong. From there the virus jetted round the world infecting over 8,000 people in twenty-seven countries. Eight hundred people died before the virus was eventually brought under control four months later.

The whole alarming episode began in November 2002 with an outbreak of an untreatable 'atypical pneumonia' in the city of Foshan in Guangdong province, China, and by January 2003 similar cases were turning up in Guangdong's capital city, Guangzhou. The virus was probably transported there by a travelling seafood merchant who was admitted to the city's hospital and sparked a major outbreak there. By the time the World Health Organisation (WHO) got to hear about it just three months after the start of the epidemic, there had been 302 cases and at least five deaths—too late to stop it snowballing out of control.

At first the microbe spread only locally within China, but in February 2003, after a sixty-five-year-old doctor who worked in a hospital in Guangzhou arrived in Hong Kong to attend a wedding, it went global. The doctor checked into room 911 on the ninth floor of the Metropole Hotel and by the time he was admitted to hospital twenty-four hours later he had infected at least seventeen others in the hotel. These people then departed to their various destinations, carrying the virus with them to five separate countries and spawning major epidemics in Vietnam, Singapore and Canada. The chain of infection widened as they passed the virus on in hospitals, clinics, hotels, workplaces, homes, trains, taxis and aeroplanes; a single virus-carrying passenger on one flight infected twenty-two of the 119 other travellers.

SARS (severe acute respiratory syndrome) begins with a flu-like illness, but instead of improving after a week or so it progresses to pneumonia. Sufferers feel feverish and increasingly breathless, with a persistent cough, as the virus colonizes the air sacs of the lungs, damaging their delicate lining and filling them with fluid. By the time they seek medical help many are fighting for breath and are immediately shipped to intensive care units for mechanical ventilation. The cough generates a spray of tiny virus-laden mucus droplets, so anyone in the vicinity is in danger of picking up the infection. Family members are at high risk; and before the danger was appreciated many health-care workers, were among the casualties, after struggling to save lives by clearing airways, artificially ventilating and resuscitating patients.

A young Hong Kong resident who visited a friend at the Metropole Hotel on the day the SARS-infected doctor was in residence was later admitted to Hong Kong's Prince of Wales Hospital, where he started an outbreak among doctors, nurses, students, patients, visitors and relatives that eventually resulted in 100 cases. One of these carried the virus to Amoy Gardens, a private housing estate in Hong Kong, where it spread like wildfire. Over 300 people on the estate caught the infection and forty-two died. Although SARS is mainly spread by coughing, the virus is excreted in faeces and since most SARS victims develop watery diarrhoea this is another possible route of infection. In fact diarrhoea was a prominent feature among the Amoy Gardens SARS victims, and some experts think the unprecedented attack rate there was caused by a partially blocked sewage system which, combined with strong exhaust fans in the toilets, created a rising plume of contaminated warm air in the airshaft that spread to living quarters throughout the building.[1] Thus the epidemic in Hong Kong took off, infecting some 1,755 people before it was brought under control (Figure 0.2).

Meanwhile the virus arrived in the US and Canada, seeded directly from the Metropole Hotel in Hong Kong. And although it did not spread in the US, the virus took hold in Toronto before doctors realized what was happening. Six of the first ten cases were from the family of an elderly couple who stayed at the Metropole Hotel (ninth floor) while visiting their son in Hong Kong. Their family doctor became the seventh victim and although she recovered, an elderly man who happened to be in the hospital emergency department at the same time as one of the family caught the virus and died.[2] The microbe then moved out into the Greater Toronto area, infecting 438 and killing forty-three before its spread was finally halted.

Dr Carlo Urbani, a WHO infectious disease expert working with a team from Medecins sans Frontières at the French Hospital in Hanoi, Vietnam, was among the first to recognize SARS as a new and dangerous infection and to note its high rate among health-care workers, who accounted for thirty of the first sixty cases in Hanoi. By warning the world of its dangers they ensured instigation of the necessary precautions worldwide, but sadly it was too late for Dr Urbani. He felt the ominous symptoms developing during a flight from Hanoi to Bangkok and alerted the authorities

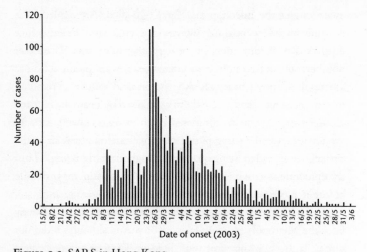

Figure o.2 SARS in Hong Kong

Source: I. T. S. Yu and J. J. Y. Sung, 'The Epidemiology of the Outbreak of Severe Acute Respiratory Syndrome (SARS) in Hong Kong – What We Know and What We Don't', *Epidemiology and Infection,* vol. 132 (2005) (Cambridge University Press, 2005), pp 781–6.

on his arrival. He battled with the virus for eighteen days in a makeshift isolation room in Bangkok hospital but died at the end of March.[3] Five of his colleagues also fell victim to the disease.

WHO's global health alert issued on 12 March caused long-unused traditional public health measures to swing into action. These included routine isolation of SARS cases and quarantine for anyone who had contacted a case to prevent spread in hospitals, while travel restrictions with country entry and exit screening were imposed to interrupt the microbe's spread in the community. These precautions, along with a high-profile media awareness campaign, brought the pandemic under control by July 2003. But before the whole episode was over there was a final sting in the tail. In late 2003 the virus jumped to two laboratory workers, one in Singapore, the other in Taiwan, while they were handling it. Fortunately these infections were not fatal and there was no further spread, but then in spring 2004 two more laboratory workers, this time in Beijing, developed SARS, precipitating an outbreak of six further cases and one death.

By the end of the pandemic there had been over 8,000 SARS cases and 800 deaths involving thirty-two countries. Worst affected was China with two thirds of cases and one third of deaths. Despite the death toll the whole episode must be regarded as a victory for those who worked so hard to contain the microbe; it could have been a lot worse. As it was, it cost an estimated 140 billion US dollars, mostly from reduced travel to, and investment in, Asia.

In contrast to the rather medieval-sounding quarantine measures that were needed to curtail the spread of the SARS virus, the search for the culprit used twenty-first-century molecular technology and was accomplished with amazing speed. A coronavirus (so-called because of its crown-like structure) was identified in SARS victims by the end of March 2003 and confirmed to be the cause by the middle of April, just two months after the doctor in Hong Kong initiated its global spread.

These days a completely new human microbe like the SARS coronavirus is most likely to be a *zoonosis*—an animal microbe that has jumped from its natural host to humans. And since more than a third of the early SARS sufferers in Guangdong were food or animal handlers, scientists hunting for its origin headed for the wet markets of Guangdong, where wild animals are sold live for the table. Armed with molecular probes they found a SARS-like coronavirus which was virtually identical to the pandemic virus strain in several species, but most often in the Himalayan masked palm civet cat, a member of the mongoose family, which is farmed in the area.[4] Fortunately these animals are not very widespread in the wild, but many experts suspect they are not the virus's natural host. They could just have acted as a go-between, picking the virus up from an unknown wild animal and passing it on to humans; so the natural reservoir of the virus in the wild is still uncertain.

Blood tests show evidence of past SARS infection in 13 per cent of Guangdong wet-market traders and animal handlers,[5] indicating that the coronavirus has jumped to humans living in this area before, and suggesting that it is likely to do so again. Indeed four new cases appeared in China in January 2004, and although they were relatively mild and did not spread further, they are a reminder that the virus is still out there waiting for another opportunity to pounce.

SARS was the first pandemic microbe of the twenty-first century, but it will certainly not be the last. Ever since HIV emerged over thirty years ago we have witnessed increasing numbers of new microbes, which are now hitting us at an average of one a year. While the SARS pandemic may be a preview of what is to come, it also gives us a glimpse of what our ancestors suffered over thousands of years: unpredictable epidemics caused by lethal microbes appearing out of the blue, killing indiscriminately and spreading fear and panic. We were fortunate in knowing how to stop SARS, but, as this book illustrates, our predecessors were not so lucky and the consequences were sometimes devastating. In later chapters we will look at well-known epidemic microbes such as bubonic plague and smallpox, as well as lesser-known killers such as the trypanosome and schistosome parasites. We will see how and why these and other microbes rose to prominence at different stages in our cultural history and the profound effect they had on the lives of our ancestors. But first, back to the dawn of time to track the origin and evolution of killer microbes, to see how they spread and invade our bodies, and how our immune system responds to the challenge.

1

HOW IT ALL BEGAN

When our solar system first formed some 4.6 billion years ago planet Earth was a very unfriendly place. Rather like the planet Venus today, Earth was so hot that carbon dioxide gas bubbled from molten rock and filled the atmosphere, causing such a massive greenhouse effect that the planet literally boiled dry. No living organism could survive under those conditions. But when Earth had cooled sufficiently for water vapour to liquefy just under 4 billion years ago, life appeared on the planet. This was not life as we know it today, but molecules that could replicate to produce daughter molecules with inherited characteristics. Darwinian evolution was set in motion and eventually microscopic single-celled organisms evolved.

These early life forms had to withstand Earth's highly volatile atmosphere with toxic gases spewing from erupting volcanoes, dramatic electrical storms and the sun's unscreened ultraviolet rays all promoting uncontrolled electrochemical and photochemical reactions. The microbes around at this time probably resembled today's 'extremophiles', a type of bacteria so-called because they thrive in all the particularly hostile corners of the globe. Extremophiles inhabit acid lakes, hyper-saline salt marshes and the superheated water issuing from hot vents at the bottom of the deepest ocean trenches where they survive temperatures up to 115°C and 250 atmospheres of pressure. They lie buried 4 kilometres deep in the polar ice caps, and lurk in rocks up to 10 kilometres below ground. Indeed it is possible that life began with microbes in rocks deep underground, where the heat is intense and there is an ample supply of water and chemicals to get the whole process started.

Extremophiles often congregate in coral-like structures called stromatolites, also known as microbial 'mats' because from the outside they look like doormats; flat, brown and hairy. These are home to thriving communities of interdependent microbes, each utilizing another's waste to produce energy in a self-sustaining food chain or micro-ecosystem. Today, microbial mats can still be seen in corners of the world such as Yellowstone Park, Wyoming, USA, lakes fed by ancient aquifers in North Mexico, and along the shores of Western Australia, where the water is rich in chemicals and undisturbed by other forms of life. Ancient layered rock structures found in these places are thought to represent the fossilized remains of stromatolites that dominated aquatic ecosystems in the Archean eon (2.5 billion to 4 billion years ago).

For around 3 billion years bacteria had Earth all to themselves and they diversified to occupy every possible niche. At this stage there was no oxygen in the atmosphere so they evolved many different ways of unlocking the energy bound up in rocks, utilizing chemical compounds of sulphur, nitrogen and iron. Then around 2.7 billion years ago a group of innovative microbes called the cyanobacteria (previously called blue-green algae) learnt the trick of photosynthesis, using sunlight to convert carbon dioxide and water into energy-rich carbohydrates. As a result, oxygen, a waste product of this reaction, slowly accumulated in

Earth's atmosphere. At first oxygen was poisonous to early life forms, but then other ingenious bacteria discovered that it could also be used to generate energy. These new energy sources were rich enough to support more complex life forms, but the emergence of multicellular organisms had to await the evolution of eukaryotic cells.

Bacteria are prokaryotes, meaning that their cells are smaller than those of all higher organisms (eukaryotes) and have a simpler structure, lacking a well-defined nucleus. But around 2 billion years ago a group of free-living photosynthetic cyanobacteria took up residence inside other primitive single-celled organisms to form the energy-generating chloroplast of the first plant cells. And in a similarly extraordinary manoeuvre oxygen-utilizing microbes called alpha-proteobacteria became incorporated into other microbes as mitochondria, the powerhouse of animal cells.

So finally, a mere 600 million years ago, the stage was set for the evolution of multicellular organisms made up of eukaryotic cells, and eventually the emergence of the plants and animals we know today. But compared to the diversity of bacteria, all other life forms, however different they may seem, are homogeneous, locked into the same biochemical cycle for energy production, and requiring sunlight for plant photosynthesis to generate the oxygen used by animals for respiration. We still rely on bacteria (in the form of chloroplasts and mitochondria) for these reactions, and on free-living bacteria for all other chemical processes needed to maintain the stability of the planet. These bacteria recycle the elements which are essential for life on Earth and are at the heart of our balanced ecosystems, those complex interdependent relationships that exist between plants, animals and the environment.

Although bacteria were the first to inhabit Earth they are not the only microbes. Single-celled protozoa, including the plasmodium that causes malaria, probably represent the earliest and simplest forms of animal life, while the tiniest of all microbes, viruses, probably also evolved several million years ago. They have diversified to infect all living things including bacteria, but exactly how and when they came into being is unknown. The genetic material of viruses consists of either DNA or RNA, but most only code for up to 200 proteins and cannot survive on their own. So viruses are obligate parasites and only when they have sabotaged their host's cells do they spring to life. Once inside they turn the cell into a factory for virus production and within hours thousands of new viruses are ready to infect more cells or seek another host to colonize.

Perhaps because they are so small, nowadays microbes seem to be overshadowed by larger forms of life, but they are still by far the most abundant on the planet, constituting some twenty-five times the total biomass of all animal life. There are well over a million different types, mostly harmless environmental microbes. They are in the air we breathe, the water we drink and the food we eat—and when we die they set about deconstructing us. Each ton of soil contains more than 10,000,000,000,000,000 (10^{16}) microbes,[1] many of which are employed in breaking down organic material to generate essential nitrates for plants to utilize; every year nitrogen-fixing bacteria recycle 140 million tons of atmospheric nitrogen back into the soil.

Bacteria and viruses are also a key part of marine ecosystems, forming by far the largest biomass in the oceans. There are at least a million bacteria in every millilitre of seawater, most abundant in estuarine waters where they break down organic matter. Marine viruses control the numbers of these bacteria by infecting and killing them, particularly when they undergo a population explosion and produce algal blooms. In coastal waters viruses greatly out-number bacteria, reaching concentrations of around 100 million in every millilitre, and totaling an incredible 4×10^{30} in the oceans. Tiny as they are, if placed end to end they would stretch for 10 million light years, or 100 times across the galaxy.[2]

As free-living organisms, bacteria have all the cellular machinery they need to grow and divide independently. They are between 1 and 10 microns in length and most contain a single chromosome. When stretched out, this coiled circular molecule of DNA reaches to about 1 mm and carries up to 8,000 genes which code for all the proteins bacteria need to survive independently of other life forms. Bacteria reproduce by binary fission, which involves making a copy of their chromosomal DNA and then simply splitting in two. *Vibrio cholerae*, as one of the fastest growing bacteria, can accomplish this feat once every thirteen minutes, and even the slowest growers like *Mycobacterium leprae* can double their numbers every fourteen days. Given ideal conditions a single bacterium could produce a colony weighing more than the Earth in just three days;[3] but fortunately conditions would be far from ideal long before that!

Bacteria are masters at survival, and when adverse conditions come along they are generally ready. Adaptability is the key to their success, yet in theory reproducing by binary fission yields offspring that are all identical to the parent—a process that apparently leaves no room for variability. But although their DNA copying machinery is accurate, mistakes occur which are corrected by a cellular proofreading system. Even so, occasional errors slip through un-noticed and these heritable changes to the genetic code (mutations) may cause changes to their offspring. This is the basis of evolution by natural selection. In humans and other animals evolutionary change is a slow process because of our long generation times, but for bacteria, which reproduce very fast and have a less effective

DNA proofreading system, rapid change by mutation is their lifeline. A single bacterial gene mutates at a rate of one change per 10^4–10^9 cell divisions, so in a rapidly dividing colony many thousands of mutants are thrown up. A few of these mutations will confer a survival advantage and these progeny will then quickly out-compete their rivals and come to dominate the population.

Bacteria have several other tricks to help them adapt rapidly to a changing environment, mostly involving gene swapping. Many bacteria contain *plasmids*, circular DNA molecules that live inside the bacterial cell but are separate from the chromosome and divide independently. They supply their host bacteria with extra survival information and can pass directly from one bacterium to another during conjugation. This involves the outgrowth of a filament called a 'sex pilus' which acts like a temporary bridge between the donor (male) and the neighbouring recipient (female) bacterium giving plasmids free access and allowing survival genes to spread rapidly through bacterial communities. Several genes that code for antibiotic resistance, allowing bacteria to survive in the face of antibiotic treatment, are carried on plasmids, and they have succeeded in spreading worldwide.

Another way that genes can jump between bacteria is by using viruses called bacteriophages, or phages for short. All viruses are cellular parasites, and phages commandeer the bacteria's protein-making machinery to generate thousands of their own offspring, most of which carry a copy of DNA identical to the parent phage. But around one phage in a million mistakenly picks up an extra piece of DNA, either from the bacterial chromosome or from a resident plasmid, and carries it to the next bacterium it infects. If this extra piece of DNA codes for a protein that improves survival then natural selection will ensure that the offspring of the recipient bacterium will prosper at the expense of others.

Sometimes phages set up long-term symbiotic relationships with their host bacteria, with the phage being safely housed inside the bacterium and the bacterium in turn being protected from infection by other more destructive phages. Remarkably, the toxin that can fatally damage the heart and nerves during a diphtheria infection, and another that causes the catastrophic diarrhoea of cholera, are both coded for by phages resident in the bacteria rather than by the bacteria themselves. Without their phages *Corynebacterium diphtheriae* and *Vibrio cholerae* are harmless.

At some stage in the distant past, groups of resourceful microbes found a niche in or on the bodies of other living things and evolved to parasitize host species. From that time on the struggle for survival has shaped the evolution of both parties. On occasion, a comfortable symbiotic relationship developed, like, for example, the microbial communities that form self-sustaining ecosystems in the guts of their hosts. For ruminants such as cows the advantages of this partnership are obvious; the microbes are bathed in nutrients and protected from the outside world while they digest the cellulose in plant cell walls which cattle are unable to do for themselves. In humans, however, the function of gut microbes is not so clear. We each house up to 10^{14} microbes, weighing in total around 1 kg, and outnumbering our own body cells by ten to one. So far, more than 400 different species have been identified which probably protect us from attack by more virulent microbes, aid our digestion and stimulate our immunity.[4] They are harmless as long as we are healthy, but if they manage to invade our tissues, perhaps through a surgical wound, they can cause nasty infections.

Of the million or so microbes in existence, only 1,415 are known to cause disease in humans.[5] But despite their significance to us, these pathogenic microbes are not primarily concerned with making us ill.

The sometimes devastating symptoms they produce are really just a side-effect of their life cycle being enacted inside our bodies. However, they certainly use each step of the infection process to their own advantage, and natural selection ensures the microbes that induce disease patterns that are best designed to assist their reproduction and spread survive at the expense of their more sluggish siblings. So over time disease patterns have been sharply honed by evolution to ensure the survival of the causative microbes. A highly virulent lifestyle, killing the victim outright, is not advantageous to microbes as they will then be without a home and probably die along with their host. Yet less virulent microbes risk being rapidly conquered by the host's immune system, and this also curtails their spread. Over centuries of coexistence of microbes and their human host, evolution has fine-tuned the balance between these two extremes to optimize survival of both species, but the rapid adaptability of microbes means that they are generally one step ahead in the ongoing struggle.

Microbe spread

An airborne microbe benefits from a host who is well enough to continue their daily routine and so keep its chain of infection alive by passing microbe offspring on to other susceptible hosts. So while only causing us mild discomfort, common cold-like viruses colonize our nasal passages, making our noses run and tickling local nerve endings to trigger our sneezing reflex. This ingenious strategy produces a fine spray of tiny virus-bearing droplets which remain airborne for long enough to infect a whole classroom, bus or train carriage full of people. And even if other airborne microbes like flu and measles viruses eventually confine us to bed, thousands of microbes have been shed by the host during the incubation period, well before the illness takes hold.

Microbes that cause gastroenteritis have found a very efficient way of travelling from one victim to another through faecal contamination of food and drinking water. While reproducing in our gut rotaviruses kill the lining cells, producing large raw areas which can neither absorb nor retain fluids. So body fluids leaking into the gut mix with dietary liquids, causing the profuse watery diarrhoea we all know and dread. This efficiently flushes the viruses back into the environment, and with each gram of faeces containing around 10^9 of them it is not surprising that the microbe easily finds another host, particularly in developing countries where around a billion people still have no access to clean water.

Some microbes are too fragile to survive in the outside world for long and so have to pass directly from one person to another. One of these, the infamous Ebola virus which occasionally finds its way into the human population from an unknown animal host, causes epidemics of a highly lethal haemorrhagic fever. The virus punches holes in capillaries and blood teeming with viruses oozes into tissues and body fluids. So while the patient is prostrate with high fever, severe pain, generalized bleeding and catastrophic vomiting and diarrhoea, the viruses in body fluids take the opportunity to pass to unsuspecting family members and hospital staff. Other directly transmitted microbes, such as those that cause syphilis and gonorrhoea, have found a niche in the warm moist surroundings of the human genital tract, exploiting the basic human instinct for procreation as a highway between victims.

Many highly successful microbes use living vectors to ferry them between hosts. Often a biting insect will oblige by ingesting the microbe while taking a blood meal from one victim and then injecting it into the next. Vector-transmitted microbes often have complicated life cycles with essential steps in the vector and the host, so both influence the evolution of the parasite. The life cycle of the malaria parasite in humans has evolved to maximize its chance of being picked up by an *Anopheles* mosquito, the only insect that can transmit the disease. The parasite colonizes red blood cells, feeding on the oxygen-carrying protein, haemoglobin. And when, forty-eight or seventy-two hours later (depending on the type of malaria parasite), the cells burst open releasing a new batch of parasites into the bloodstream, the waste products from the parasite's meal trigger the high fever, rigors and malaise characteristic of a malaria attack. These symptoms are severe enough to keep the sufferer lying still so that a grazing mosquito can take a full blood meal undisturbed. So in this case synchronizing the debilitating symptoms with the release of large numbers of new blood parasites greatly enhances the microbe's chance of survival.

Malaria and several other vector-transmitted microbes are restricted to tropical regions by their vectors, which require high temperature and rainfall to breed. But microbes which are not so fussy about the vectors they use can spread further afield. The mosquito-transmitted West Nile fever virus, which can cause fatal encephalitis, recently crossed the Atlantic to hit New York in 1999. Its usual homelands are in Africa, Asia, Europe and Australia, where it uses a whole range of mosquito vectors. The virus is primarily a bird microbe but humans can get infected when bitten by a virus-carrying mosquito. By exploiting the virgin populations of birds and humans in the US the virus moved in a wave across the continent, reaching the west coast, the Caribbean and Mexico in just four years (Figure 1.1). Already it has colonized 284 US bird species and utilized fifty-eight types of mosquito as vectors. Clearly the virus has established a base in the US for the foreseeable future.[6]

Epidemics

Most pathogenic microbes constantly live on a knife edge. They are locked into a continuous chain of infection; one break in the chain and they are dead. So they must continuously jump from one susceptible host to another, infecting, reproducing and moving on before the host's immune system wipes them out. Epidemics strike whenever and wherever microbes find a large susceptible group of people to infect and can successfully forge a path between them. Given the right conditions a microbe will rip through a population, infecting and perhaps killing until it runs out of people to infect. When all are either dead or recovered and immune to further attack, the microbe will move elsewhere and only return when there are again enough susceptible people to sustain the chain of infection.

When an epidemic strikes it is epidemiologists who do the detective work to uncover the cause, predict the size of the outbreak and suggest effective control measures. A key number for them is R_0, which represents the basic reproductive rate of an epidemic; that is, the average number of new cases infected from each case in a susceptible population (Figure 1.2 and Table 1.1). It is important to know the value of R_0 when an epidemic threatens because if it is greater than one then the infection rate is increasing and an epidemic is likely. Conversely if R_0 is less than one then the infection is not self-sustaining and it will fizzle out. Monitoring this value during an epidemic (now known as R—the case reproduction number) gives an indication of how long it will last. R is generally high as an epidemic takes off and then falls as more and more people become immune to the microbe. And when it falls below one everyone can heave a sigh of relief, knowing that the worst is over.

The R_0 value for a microbe encapsulates the whole chain of events that make up its life cycle from penetrating and invading

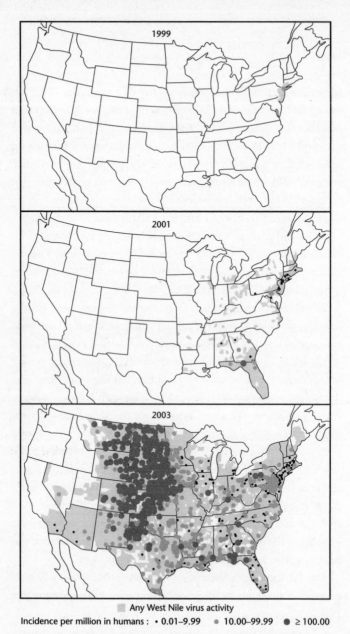

Figure 1.1 Incidence of West Nile fever in the USA, 1999–2004

Source: L. R. Petersen and E. B. Hayes, 'Spread of West Nile Fever in the US 1999–2004', *New England Journal of Medicine* 351(2004): 2258.

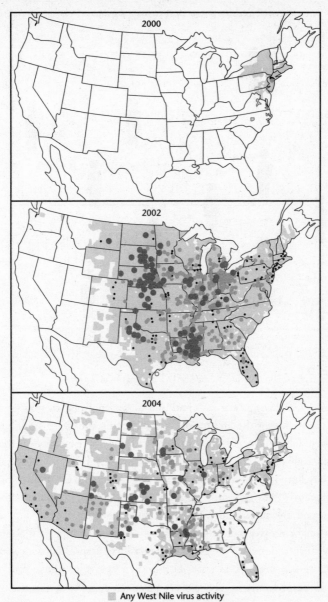

2000

2002

2004

Any West Nile virus activity

Incidence per million in humans : • 0.01–9.99　• 10.00–99.99　● ≥ 100.00

UHB TRUST LIBRARY QEHB

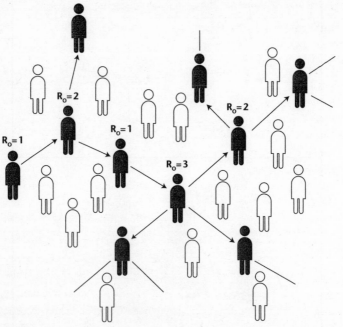

Figure 1.2 R_0: the basic reproductive number of an epidemic

a host, infecting host cells and producing offspring, to finding a way back into the environment and locating another susceptible host. The success of these manoeuvres depends not only on the microbe itself, but also on its host population and the environment they both live in. So the dynamics of an epidemic are defined by the microbe's transmission route, the length of its incubation period, the size and density of the susceptible population, and, if a vector is involved, its geographic range. For example, although sexually transmitted disease (STD) microbes can spread as widely as those that are air- or water-borne, their epidemics move much more slowly and involve a more restricted population. Classically

UHB TRUST LIBRARY QEHB

Table 1.1 R_o values for human and animal microbes.

Pathogen	Host	Location	R_o
TB	humans	Europe	4–5
Bovine TB	possums	New Zealand	1.8–2.0
Bovine TB	badgers	UK	2.5–10
Rabies	foxes	Belgium	2–5
FMDV	buffalo	South Africa	5
FMDV	cattle	Saudi Arabia	2–73
Smallpox	humans	UK	3–11
Leishmania	dogs	Malta	11
HIV	humans	East Africa	10–12
AHSV	zebra	South Africa	31–68
Trypanosomes	cattle	West Africa	64–388

TB micobacteria tuberculosis FMDV foot and mouth disease virus HIV human immunodeficiency virus AHSV African Horse sickness virus

STD epidemics begin in young adults and target the most sexually active. For R_o to exceed one on average each case must infect more than one other, but in practice 20 per cent of those infected account for 80 per cent of the spread. These 20 per cent are 'super-spreaders' at the hub of large sexual networks, be they commercial sex workers or promiscuous gay men. HIV, greatly assisted by its long silent incubation period averaging eight to ten years, encircled the globe by using these networks before it was even recognized. Indeed the epidemic among gay men in the US in the early 1980s was thought to be sparked by one super-spreader, 'case zero'—a city-hopping, gay airline steward who had sexual links with at least forty of the earliest cases in ten different cities.[7]

In all epidemics victims display a spectrum of disease severity varying from fatal cases on the one hand to mild disease on the other, and in most outbreaks there are also silent infections with the microbe colonizing its host without causing any disease at all. Sometimes these silent infections form a sizeable proportion of the total; in some flu epidemics, for example, up to twice as many people are infected as suffer any illness, and with polio less than one in a hundred of those infected come down with paralytic disease. But silently infected individuals are a major source of spread to others since they remain well and are unaware that they are infectious. So, in order to calculate R_0 and get an accurate picture of the size and progress of an epidemic, these silent infections must be detected by laboratory tests and taken into account.

SARS coronavirus, being new to humans, has had little time to evolve with its new host and is not well adapted for spreading in the human population. It produces severe disease, killing 10 per cent of its victims, and is spread in heavy mucus droplets which can only travel a short distance, infecting close contacts. SARS victims are not infectious during the incubation period and do not shed the virus after recovery. Add to this the fact that during the pandemic there were virtually no silent infections and it is not surprising that R_0 is in the modest range of 2–4. But despite all these disadvantages the virus spread round the globe, aided by super-spreaders (like the doctor at the Metropole Hotel in Hong Kong) and fast international air travel. It is interesting to imagine how the virus would have fared if it had emerged hundreds of years ago before we had any knowledge of how to stop it spreading. Would it have remained a major international killer like smallpox or would it have evolved into a milder form and joined the band of viruses that cause flu-like illnesses today?

Host Resistance

The battle between humans and microbes has been raging ever since *Homo sapiens* evolved, and was presumably ongoing in our primate ancestors long before that. From a present-day standpoint it is hard to see how humans have been able to compete with the apparent ingenuity of microbes. But the story is one of co-evolution over thousands of years, and its history is written in our genes. Each time an infectious disease hit our ancestors it weeded out the most susceptible, leaving only the more resistant survivors to pass on their genes to future generations. Thus, step by step, the population slowly built up genetic resistance to the whole range of pathogenic microbes, while at the same time many microbes evolved to be less virulent. Consequently most infectious diseases became less severe over time. Now we are all descended from a long line of fore-bears who survived epidemics and spawned offspring with inbuilt resistance, and it is thanks to them that we are here to tell the tale.

The evolution of resistance is perhaps most clearly illustrated by the fascinating association between malaria and the inherited blood diseases, thalassaemia and sickle-cell anaemia. These muta-tions cause red blood cell disorders which are fatal in homozygotes (those who inherit mutated genes from both parents) if untreated. They should therefore have died out over time, but they have been retained in the human gene pool because they protect heterozygote carriers (with one mutated gene) against death from malaria. Over centuries thalassaemia and sickle-cell anaemia carriers survived while many others died of malaria, and the frequency of these genes gradually increased until now they are amazingly common among people living in areas where malaria is or was endemic. Today up to 40 per cent of sub-Saharan Africans carry the sickle-cell anaemia gene and 70 per cent of the Papua New Guinean population are carriers of various thalassaemia mutations.

There must be many similar undiscovered genes that enable resistance in the human genome, most of them probably coding for proteins which enhance our immune response, having been forced to prominence by the relentless onslaught of killer microbes such as smallpox, plague and diphtheria. By constantly challenging our immune system these microbes have helped shape it into a highly complex and sophisticated fighting machine, with a fast response mode to stall new invaders and a memory to prevent re-invasion.

White blood cells are the mainstay of our immune system. There are many types of these mobile cells, including polymorphs, macrophages and lymphocytes, which travel in blood and patrol the tissues hunting for invading microbes and stopping them in their tracks. When a microbe succeeds in breaching our defences polymorphs and macrophages are first on the scene. They secrete chemicals called cytokines which increase the local blood flow, opening up the region to other immune cells. This produces the red painful swelling typical of an inflamed area, and also the non-specific flu-like symptoms (fever, head and muscle aches, and lethargy) we experience at the start of most infections. Macrophages are large amoeba-like cells that engulf and digest invading microbes. Then, laden with microbial proteins, they make their way to local lymph glands where they interact with lymphocytes to initiate the later, more specific phase of the immune response.

Lymphocytes are small innocuous-looking cells but they make up a formidable army which defends us against all comers. The body contains an incredible 3×10^9 lymphocytes, each with its own receptor that fits just one segment of a foreign protein. They circulate in blood and congregate in lymph glands, and when they meet a macrophage carrying a bit of foreign protein that fits their particular receptor they are galvanized into action, growing and dividing to form a whole army of cloned offspring ready to deal with the invading microbe. B-lymphocytes produce antibodies which mop up bacteria while T-lymphocytes are armed with chemicals which punch lethal holes in virus-infected cells. With these immune mechanisms in full swing most microbes are routed, but a few have evolved ingenious ways of evading the attack or tricking the immune system into thinking they are part of the body's make-up. For example, *Mycobacterium tuberculosis* bacteria survive by living in macrophages which can engulf but not destroy them, while herpes viruses hide inside long-lived cells, often without expressing any proteins for the immune system to target. These silent persistent microbes may emerge years later if the body's immunity is suppressed. On the other hand, protozoa such as malaria and trypanosome parasites can change the protein composition of their coat throughout their life cycle so keeping one step ahead of the immune system, and HIV achieves the same end by mutating regularly.

One of the most intriguing things about the immune system is its ability to remember past encounters with microbes and so prevent reinfection by the same microbe. This is achieved by retaining a few microbe-specific memory lymphocyte clones after an infection is cleared, so that they are ready to respond rapidly at the next encounter. This prevents the microbe from getting a foothold the second time round and explains why as a rule we only get one attack of each acute infectious disease like measles and mumps in a lifetime.

This immunological memory is the principle underpinning vaccination. Today we regularly induce immunity artificially by giving a dose of a killed or weakened microbe which tricks the body into responding, as if to the natural infection. This generally gives lifelong protection and prevents epidemics by interrupting the microbe's life cycle. But for most of our history the natural cycles of infections held sway, and in the rest of this book we look at their impact on human history.

2

OUR MICROBIAL INHERITANCE

Homo sapiens and his closest living relatives, the great apes (gorillas, chimpanzees and bonobos), split from their common ancestor in Africa around 6–7 million years ago. Only a few scattered fossil remains are available to provide snapshots of his subsequent evolution through a series of hominids with progressively more erect posture, increased brain size, loss of body hair and improved manual dexterity. One such hominid, *Homo erectus*, recognized in the fossil records from around 250,000 to 1.8 million years ago, abandoned the rainforest and began to hunt on the open plains of East Africa. This move, perhaps prompted by a change to a drier climate that increased the savannah area at the expense of the forest, eventually led to their migration out of Africa, so that by around 1.7 million years ago these creatures had spread as far as Indonesia, China and Europe. For thousands of years these hominids roamed the land in small bands, gathering fruits, leaves and roots, and using crude stone tools to hunt for small prey.

All animal species have their own collection of parasites that have co-evolved with them over many centuries, and our ape-like ancestors were no exception. They and their parasites were part of the balanced ecosystem of the tropical African rainforests, and as long as the situation remained stable host and parasite could continue to evolve together, living in a state of mutual coexistence which caused little problem to the host. At this stage it is impossible to say exactly what these parasites were, but although they may have weakened heavily infected individuals, they are unlikely to have been fatal.

Modern humans probably evolved in Africa between 150,000 and 200,000 years ago, and a subsequent exodus 50,000–100,000 years ago eventually populated the whole of the modern world. These Cro-Magnon people are our true hunter-gatherer ancestors, and as exemplified by their famous paintings in the Lascaux caves in the South of France, they were technologically and socially more advanced than their predecessors. They fashioned clothes and shelters from animal skins to keep out the cold, and made sophisticated hunting tools with which they could tackle large game without fear of being preyed upon themselves. For the first time humans topped the food chain.

These early hunter-gatherers are sometimes envied for their egalitarian society and their lifestyle is idealized as being 'in tune with nature'. On the other hand the seventeenth-century English political philosopher, Thomas Hobbes, described 'life in the state of nature' as a 'war of all against all' and the hunter-gatherer life as 'solitary, poor, nasty, brutish and short'. This chapter looks at the truth behind these differing opinions and considers the impact of microbes on the lives of individuals and the population as a whole.

As the name suggests, hunter-gatherers are nomads who live in small groups or bands, constantly occupied by the pursuit of food. They move with the seasons, following herds and crop cycles— hunting, trapping, fishing and gathering wild fruits, roots, leaves and seeds. This was man's routine occupation for thousands of years before it was almost completely replaced by the agricultural revolution that began some 10,000 years ago. However, a few hunter-gatherer tribes still survive in remote pockets of the world, or at least did so within living memory.

With no first-hand written accounts of the ancient hunter-gatherer way of life (writing was only invented around 3000 BC), we must reconstruct it as best we can with information gleaned from their settlements, cave dwellings, burial sites and skeletal remains. The few hunter-gatherer tribes that still survive today, like the Australian Aborigines, the Kalahari San people, the African Bushmen, and the Pygmies of the African rainforests, also provide some useful information, but as none of these tribes is entirely free from contact with the outside world, study of their microbes must be interpreted with caution. A promising new line of enquiry involves the use of molecular genetic probes to pick up microbe-specific DNA or RNA sequences in human remains. And although these techniques are still being developed, they are already proving to be highly sensitive tools for gaining new insights into the ancient history of microbes and pinpointing when and where they first infected humans.

A typical hunter-gatherer band contained thirty to fifty people, generally a few extended family groups, and formed part of a loose network of bands that would meet from time to time perhaps to celebrate a marriage or bury their dead, when members would take the opportunity to exchange information. Each band had a defined territory and the size of the band was dictated by the availability of food in their particular patch. On average, hunter-gatherers required around 1 square mile of foraging area per person, so the number of people in a band was critical: past a certain tipping point further increase would be self-defeating since it would mean travelling further for food, and, with no form of transport, carrying heavier loads back to their settlement. So every now and then a growing band would split in two with half moving to a new territory.

Hunter-gatherer bands were small enough for social and political structures to be simple and informal, most business being conducted on a personal level. And since virtually everyone was engaged in gathering and preparing food, all had the same rights to resources. There was little need for social class distinctions, and the society was generally mutually supportive. But implicit in the hunter-gatherer lifestyle are regular moves from one settlement to another every few days, weeks or months, depending on the availability of food in the foraging area. So although this lifestyle suited fit adults, there was little support for the sick, elderly and infirm, and archaeological remains suggest that they were sometimes abandoned. Similarly, large young families would have hindered the mobility of the band, and there is evidence that infanticide was commonly practised to control family size, giving an average gap between children in a family of four years.

Generally, hunter-gatherers seem to have been reasonably healthy. In both modern and ancient bands members were generally lean and fit, and although they probably experienced occasional food shortages, on the whole they were adequately nourished. Their life expectancy was around twenty-five to thirty years, with infant mortality between 150–250 per 1,000 births.[1] These figures may seem high when compared to present-day values in the West (infant mortality in Western societies is 3–10 per 1,000, and a life expectancy is over seventy years), but they compare favourably with values from any time in history up to the eighteenth or nineteenth century, and are almost equivalent to the lowest life expectancy and highest infant mortality figures from developing countries today.

Skeletal remains clearly show that hunter-gatherers did not generally die of starvation, dietary deficiencies or injury, but unfortunately bones on their own are not very useful for providing evidence of infectious diseases since microbes do not generally leave fossil evidence of their presence. Only the few microbes that attack bones and joints, such as those causing tuberculosis, syphilis and leprosy, can be diagnosed with any certainty, and none of these was common among ancient hunting bands. Despite this lack of evidence many experts believe that infectious diseases were among the commonest causes of death in hunter-gatherers, but the nature of these diseases can only be guessed at by using our knowledge of present-day infections and the insight into their origins afforded by modern molecular techniques.

Today microbes use almost every conceivable transmission route to bridge the gap between susceptible hosts, but many of these routes were barred to microbes during the hunter-gatherer era because of the small size, isolation and mobility of the bands. It is thought that the airborne microbes that cause the classic acute childhood epidemics of modern times did not exist in Palaeolithic times. Once seeded into a band these microbes would have ripped through its members with no trouble at all, but the sparse population of small bands, which only met up occasionally and probably foraged in areas many days' walk apart, would have prevented further spread. So microbes would soon run out of susceptible people and be unable to maintain their vital chains of infection. Indeed, studies on acute infections like measles, mumps and whooping cough in relatively isolated modern South American tribes show exactly this pattern.[2] When one of these microbes reaches a band from the outside world it sweeps through its members but fails to spread to all bands in the region and cannot gain a permanent foothold.

However, the more recent observation of the devastating effects the acute childhood infections had on isolated American Indian hunter-gatherers when introduced by European explorers in the fifteenth century is probably the most compelling evidence we have to suggest that these microbes had not infected them before. They had not evolved resistance through previous encounters and they died in their thousands (see Chapter 5). Of all the acute childhood infections, chickenpox is the rare exception in that it almost certainly did infect hunter-gatherers. The causative virus, *varicella zoster*, is a member of the ancient herpesvirus group that humans inherited from their ape-like ancestors. These viruses, which also include the cold sore and genital herpes viruses, are carried by people in all the world's most remote tribes and are so well adapted to humans that they are virtually never life-threatening. They have overcome the problem of spreading among people in sparsely populated areas by dodging immune attack and setting up an association that lasts a lifetime. This silent infection reactivates intermittently, as recurrent cold sores for example, to produce new viruses which ensure their survival. So although chickenpox behaves like a classical acute childhood infection, once the acute illness is over the microbe hides in the host's nerve cells and can reappear years later to cause a bout of shingles. This nasty skin rash consists of tiny blisters filled with viruses ready to spark a chickenpox epidemic among a new generation of susceptible children.

Contamination of food and water with human sewage is a highly successful transmission route for many microbes, particularly where standards of hygiene are low. But since hunter-gatherers collected and ate their rations on a daily basis and had no facilities for storing either food or water, faecal contamination is unlikely to have been a major problem. Many of the large parasitic worms common in

Africa today are transmitted by this route and cause anaemia from chronic intestinal bleeding. The resulting lethargy could have serious consequences for a hunter-gatherer band, but since effective transmission only occurs with a build up of faecal material containing parasite cysts and eggs, their frequent moves, abandoning their waste materials along with their camps, probably protected them from heavy worm infestations. The same argument applies to microbes that rely on vectors such as rats, lice and fleas to ferry them between hosts, and those that require an intermediate host, such as the parasite that causes schistosomiasis and that needs to infect a water snail to complete its life cycle (see Chapter 3), because these would also be left behind in an abandoned camp site.

Malaria

So far it seems that hunter-gatherers conveniently avoided many of the microbes which plagued later humans, but microbes with flying vectors could perhaps have increased their range sufficiently to overcome the problem of the sparse population. The commonest of these diseases today is malaria, which infects 300–500 million people every year and kills over a million. And since Africa is the presumed original homeland of humans and is the continent worst affected by malaria today, it has long been assumed that malaria was a scourge of early humans. There is certainly evidence of malaria infection in Egyptian mummies dating from 3000 BC (see Chapter 3), and the disease is clearly described in the Chinese *Nei Ching* (*The Canon of Medicine*), written in 2700 BC, but there are no records dating back as far as Palaeolithic times. So although the disease is undoubtedly ancient, exactly how ancient is presently unknown.

Previously called 'the ague' (a diminutive of the Latin *febris acuta* meaning 'acute fever'), the name malaria (*mal aria* or 'bad air') was popularized by the Italians in the nineteenth century to describe the disease which had plagued their country for centuries. Because it was most common among people who lived and worked in the Pontine Marshes and Roman Campagna during the summer months, it was generally believed to be caused by bad air or 'miasma' rising out of the swamps.

Malaria can appear in many different guises, depending on the type of parasite and the age and level of immunity of the victim. In its acute form in the non-immune host, the parasite begins by causing a flu-like illness which settles into a characteristic cyclical pattern. As each paroxysm begins the body temperature climbs precipitously and yet sufferers feel severely chilled. With teeth-chattering rigors and uncontrollable shivering, they curl up, miserably clutching at blankets for warmth. When the fever peaks at 39–41.5°C, the exhausted patient breaks out in a profuse sweat and the temperature falls. Symptoms of drowsiness, delirium, seizures or coma indicate the onset of cerebral malaria caused by sludging of parasitized red blood cells in the blood vessels of the brain. This is invariably fatal if untreated and is the commonest cause of death in the acute phase of the disease. In cases with no brain involvement fever attacks continue for weeks or months, but gradually resolve as immunity develops. The type of malaria parasite infection determines the periodicity of the paroxysms which occur every forty-eight (tertian fever) or seventy-two (quartan fever) hours, and some parasites can set up chronic infections which relapse at intervals for many years.

In areas of intense malaria transmission young children are the worst affected as immunity to malaria builds up slowly and does not completely protect against attacks until they are four to five years old. But, although slow to accumulate, immunity is rapidly lost without constant contact with the parasite. So anyone who leaves and then returns to an endemic area, as hunter-gatherers might have done in their wanderings, is liable to severe attacks. Similarly, in areas where malaria is epidemic, arriving with the rains and disappearing in the dry season, people of all ages suffer severe bouts of the disease at the beginning of the rainy season. Nowadays the overall death rate from malaria is around 1 per cent, but this can reach 30 per cent in severe epidemics among the non-immune.

Today a million or so children die from malaria each year, but among hunter-gatherers the death of a young child was less of a threat to the survival of the band than the loss of an active adult. After all, children were not involved in collecting or preparing food, and could be more quickly replaced than adults. So it was the burden imposed by chronic malaria in adults that would be the most arduous for hunter-gatherers. As the Prefect of Girgenti in Sicily wrote of the disease in his country in 1908:

The enormous prevalence of the disease has the most serious social consequences because the infection–tenacious and lasting–undermines the body. Malaria causes physical decline . . . , it prevents growth, and it alters the very structure of the population. . . . Fever destroys the capacity to work, annihilates energy, and renders a people sluggish and indifferent. Inevitably, therefore, malaria stunts productivity, wealth and well-being.[3]

This long-term debility, affecting both the victim and the community as a whole, if widespread, would certainly have endangered the viability of hunter-gatherer bands.

Malaria is caused by a protozoan called Plasmodium (the name derived from the Greek meaning 'a matrix containing many nuclei'), a parasite with a complex life cycle in two hosts—an asexual phase in its primary vertebrate host and a sexual phase in its mosquito vector. The parasite probably evolved from a once free-living pond protozoan that had adapted to parasitize the aquatic larva of a flying insect. Over the course of time this protozoan developed its two-host life cycle, parasitizing both the bloodsucking insect as well as its victims.

The Plasmodium that causes malaria in humans was first identified in 1880 by Charles Louis Alphonse Laveran, a French army doctor who was working in Algiers when he noticed some unusual cells in the blood of soldiers suffering from malaria. These cells contained a strange dark pigment, and as he watched them under the microscope some literally swelled up and burst, releasing a dozen or so tiny microbes with furiously lashing flagella. Laveran was sure that these were the cause of malaria, but just a year before some influential Italian doctors had identified a bacterium, *Bacillus malariae,* that they claimed was the cause, so his report was greeted with some scepticism. However, in the end he managed to convince a group of parasitologists in Rome of the seriousness of his claim, and by 1884 they too had identified parasites in the blood of malaria victims. Very soon the different types of plasmodium were linked to the two-and three-day parasite life cycles, and the significance of the periodicity of the fever attacks, which coincide with the release of new parasites from infected red blood cells, was finally understood. Another twenty years passed before a British doctor, Ronald Ross, serving in the Indian Medical Service, clarified the microbe's complicated life cycle (Figure 2.1). His interest in the transmission of the malaria parasite was sparked by a visit to Sir Patrick Manson in London, a distinguished pioneer of tropical medicine who, while working in China, made the astounding discovery that the filarial worm that causes elephantiasis is carried by mosquitoes. He encouraged Ross to look for the malaria parasite in mosquitoes, and so Ross spent the next three years in Bangalore and Secunderabad trying to find plasmodia in mosquitoes fed on malaria blood. He knew little

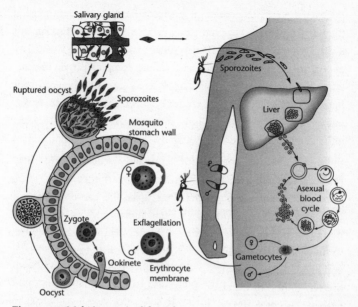

Figure 2.1 Malaria parasite life cycle

about the different mosquito species, but was finally rewarded by seeing the parasites in the stomach of mosquitoes he called 'dapple-winged', and in 1897 he reported his findings in the *British Medical Journal* in a paper entitled 'On some peculiar pigmented cells found in two mosquitoes fed on malaria blood'.[4] But before he could unravel the whole of the parasite's life cycle the army transferred him to Calcutta, where his studies were held up by a lack of malaria cases. In frustration he turned to studying malaria in birds and soon sorted out the mosquito transmission route. He found that the parasites reproduce sexually in the stomach of the mosquito vector and then move to its salivary gland, where they are poised ready for injection into the mosquito's next victim.

Once the various steps of the bird malaria cycle were unravelled the human parasite life cycle was quickly found to be similar, but it was Professor Giovanni Battista Grassi, working at the University of Rome who fitted the final piece of the complex puzzle in place when he identified the female *Anopheles* mosquito as the vector for human malaria. At the time rivalry between these researchers was intense and no one is sure how much Grassi's contribution was influenced by Ross's work; obviously the Nobel Prize committee thought it was, since both Laveran and Ross received Nobel Prizes for their work but Grassi did not.

There are nearly 400 named species of *Anopheles* mosquito, of which forty-five are important malaria vectors, and the conditions that these insects require to breed determine the global distribution of the parasite. Only the female mosquito can transmit malaria because only she sucks blood, requiring it to fuel egg production. Once she has ingested plasmodia from her victim it takes around two weeks before the parasite cycle in her stomach is complete, so she must survive at least this long after feeding in order to transmit the parasite. Suitable ambient temperature and humidity are essential; malaria transmission does not occur at temperatures below 16°C or above 30°C. All mosquitoes need access to water at the larva stage, but it is their species-specific behaviour patterns that determine whether or not they transmit malaria. *Anopheles (A.) gambiae* is the major malaria vector in Africa today and is highly adapted to its task of spreading the microbe. It is long-lived, has a particular liking for human blood and lives in and around human dwellings, breeding in the stagnant water of wells and puddles. In areas where malaria is particularly rife parasite-laden *A. gambiae* may deliver around 1,000 bites per person every year.[5]

With the help of this highly specialized, mobile injecting machine the malaria parasite might have threatened small travelling bands of hunter-gatherers in Palaeolithic times, but although it is not known for sure when the close association between *A. gambiae* and humans began, it was probably not until the dawn of agriculture in Africa around 5,000 years ago. Experts think that by clearing the forest for planting crops humans stimulated a population explosion among *Anopheles* mosquitoes that had previously lived in the occasional sun spot in the forest created by a fallen tree. When humans started living in fixed agricultural communities with relatively dense populations, these mosquitoes could at last live entirely off humans, breed in the inevitable standing water in and around their settlements, and evolve their present-day habits.

Today the Plasmodium species infects all major groups of land vertebrates, with its most ancient forms, thought to be the descendants of dinosaur malaria parasites, now infecting birds and reptiles. All other malaria parasites, including those in primates, have probably evolved from this species, diverging some 130 million years ago.[6] So our ape-like ancestors probably carried primitive malaria parasites, but most experts agree that the malaria parasites we recognize today first rose to prominence in sub-Saharan Africa and from there spread throughout Africa, across the Mediterranean, overland to Asia and Europe, and, much later, crossed the Atlantic with human migrants to the Americas (see Chapter 5). Their spread to each new area was preceded by the migration of bloodsucking *Anopheles* mosquitoes that have existed in Africa since ancient times, well before the evolution of humans. But the question of when present-day malaria parasites evolved in humans is still hotly disputed.

Of the twenty-five species of Plasmodium that parasitize primates, only four infect humans: *P. falciparum, vivax, malariae,* and *ovale,* which probably evolved separately in other animal species. Infection with *P. malariae, ovale* or *vivax* is rarely fatal, but these parasites can set up chronic infections. *P. ovale* and *vivax* can hide in the liver and reappear at intervals for two to three years after infection, and *P. malariae* can cause a lifetime of relapsing illness. *P. falciparum* on the other hand cannot set up a chronic infection, but it causes the most severe disease and is responsible for almost all malaria deaths today. This is by far the commonest type in Africa today, so what could its impact have been on early hunter-gatherer bands?

Surprisingly, despite its virulence, *P. falciparum* is not easily transmitted between humans. Its intense transmission in Africa is almost entirely dependent on *A. gambiae* because of the insect's preference for biting humans, and although the parasite maintains a presence in Papua New Guinea, Melanesia, Haiti and a few hot spots in South America, the vectors in these regions are far less efficient. Most evidence points to a West African origin for present-day *P. falciparum*, but scientists are still trying to pinpoint exactly when this occurred. They have analysed the DNA sequence of genes from parasites of humans and other species to determine how closely related they are. Using the molecular clock technique, which makes the assumption that the more genetic differences between the parasite species the longer they have evolved separately, they find that *P. falciparum* is most closely related to the chimpanzee malaria parasite, *P. reichenowi,* and estimate that the two species diverged 4–10 million years ago.[7] This date conveniently overlaps with the proposed date of divergence of hominids from chimps, suggesting that both inherited the parasite from their common primate ancestor. If this is correct then an ancestral form of *P. falciparum* must have infected our hunter-gatherer ancestors in Palaeolithic times. But most experts do not think that modern-day *P. falciparum* could have survived in humans with the inefficient mosquito vectors that were around at the time, particularly among the sparse population of hunter-gatherer bands.

Scientists are now struggling to identify when today's *P. falciparum* first emerged in humans by measuring the length of time to its most recent common ancestor, that is, the parasite from which all others evolved. Again they are using the molecular clock to assess genetic divergence, this time between *P. falciparum* strains from different parts of the world. But the results are conflicting. Some groups find a 'population bottle-neck' as recently as 5,000–10,000 years ago, suggesting that the current global *P. falciparum* population arose from a very small number, theoretically a single strain, at that time.[8] This led scientists to propose the 'malaria's Eve' hypothesis, relating the beginnings of modern-day *P. falciparum* infection to the change from the hunter-gatherer lifestyle to slash-and-burn agriculture that occurred in Africa around that time, and that would have coincided with evolution of the mosquito vector, *A. gambiae*, critical for efficient transmission of *P. falciparum*. These scientists also argue that this relatively short coexistence between man and parasite accounts for the virulence of *P. falciparum*, and estimate 5,000–10,000 years of coexistence to be sufficient for malaria-resistant genes like thalassaemia and sickle-cell anaemia to reach their present levels in malaria-exposed populations (see Chapter 1). In contrast, using the same molecular techniques, other researchers find a much more genetically diverse *P. falciparum* population that suggests the parasite has a far longer history in humans, dating the transfer to around 100,000–180,000 years ago.[9] They argue that the virulence of *P. falciparum* is not because it is a relatively new human pathogen but because, as discussed in Chapter 1, a debilitated host is advantageous for parasite transmission. But the question of whether *P. falciparum* could have maintained its human to human transmission in the sparse hunter-gatherer bands of 100,000 years ago still remains unanswered.

The molecular genetic techniques used in both these studies, although powerful, are relatively new and not without problems. There are several reasons why results from separate research groups may not agree, particularly as the two groups analysed different sets of parasite genes, and their *P. falciparum* isolates came from different parts of the world. Also the assumption that the rate of change in parasite DNA is constant over time is only correct if the genes examined are not subject to selection pressure which might induce rapid change. Since *P. falciparum* has two hosts (mosquitoes and humans) and is under constant pressure to evolve ways of counteracting the immune system of both, this assumption may not be correct.

Clearly, only analysis of more genes and further *P. falciparum* isolates can resolve the issue of when this parasite took hold in the human population, although, on balance, it seems unlikely that modern-day *P. falciparum* could have survived in sparse hunter-gatherer bands without the help of *A. gambiae* as its vector. However, before drawing any firm conclusions about malaria in hunter-gatherers we must briefly consider the other types of human malaria.

P. vivax, which is the most widespread of the three non-*falciparum* human malaria parasites, has an interesting history. This parasite is most closely related to *P. cynomolgi*, which infects monkeys in South and South-East Asia. This does not mean that human infection arose in Asia, as the two strains diverged 2–3 million years ago at a time when their ancestral hosts were probably widely spread throughout tropical and subtropical regions. In fact there is good indirect evidence that modern-day *P. vivax* was restricted to Africa until fairly recently, perhaps around 15,000 years ago, only spreading further afield as the global temperature rose at the end of the last Ice Age.[10] Evidence for this comes from genetic studies showing that 97 per cent of people from West and Central Africa are negative for a blood group called Duffy, whereas in the rest of the world virtually everyone is positive. The Duffy protein is an essential cell receptor for *P. vivax* and without it the parasite cannot infect red blood cells. Consequently, today *P. vivax* infection is very rare in West and Central Africa. The Duffy negative mutation is harmless, but only homozygotes (with two mutated genes) are resistant to *P. vivax*. When the mutation first arose the chance of one mutation-carrying person mating with another to produce homozygous Duffy negative offspring must have been vanishingly small. And even when this did occur the effect would be diluted again when most of the next generation inevitably took Duffy positive mates. Clearly it must have taken a very long time, and a very strong selection pressure, for the Duffy negative mutation to reach its present high level in Africa. So we must conclude that *P. vivax* has been around in Africa since ancient times and had a major negative influence on human survival.

Although it is not possible to pinpoint exactly when *P. vivax* first affected humans, since Duffy negative people are confined to Africa it seems fair to assume that it evolved in its present form after the major human migrations out of Africa, certainly no earlier than 100,000 years ago. And using a similar argument, since the Duffy negative mutation is not found in other areas of the world where *P. vivax* has infected people for around 5,000 years, the parasite must have been in Africa for longer than this. So, given that *P. vivax* can establish a chronic infection in humans and is therefore not dependent on a constant chain of transmission, it could have presented a problem to Paleolithic hunter-gatherers in Africa.

But perhaps the most likely candidate for infecting hunter-gatherers is *P. malariae,* a parasite that also infects chimpanzees in West Africa.

Unlike *P. ovale*, which cannot survive outside the tropics, *P. malariae* can spread in temperate as well as tropical and subtropical regions. And since this parasite can live in its host for a lifetime it is certainly the best adapted for survival in sparse, mobile hunter-gatherer bands. However, the question of whether malaria really caused major problems to hunter-gatherer bands still remains unanswered, and evidence from modern African hunter-gatherers is of little help. Pygmy tribes have a low rate of malaria, but this is because of their high incidence of the sickle-cell gene, which tells us that they have battled with *P. falciparum* for at least 5,000 years. In contrast, the San or Bushmen hunter-gatherer tribes living around the Kalahari Desert in South Africa and Botswana lack any evidence of genetic resistance to malaria and those who hunt in malaria endemic areas suffer badly from the disease, but it is difficult to relate this information to ancient hunter-gatherer bands living at a time when the efficient *A. gambiae* vector did not exist.

On balance, it seems likely that some form of malaria, probably caused by *P. malariae* and/or *P. vivax*, affected the lives of Palaeolithic humans. And although these parasites are not generally lethal, the severe lethargy caused by chronic infection would certainly have endangered the survival of hunter-gatherer bands which needed all their adult members to actively forage for food. However, it is clear that malaria did not reach its present-day levels in Africa until the agricultural revolution provided the ideal conditions for a highly efficient mosquito vector to evolve, allowing microbes to flourish.

As soon as humans began to hunt big game on the grasslands of East Africa they entered a new environment that was populated by huge herds of wild animals and the parasites they carry, and some of these unfamiliar microbes took the opportunity to jump species and infect man.

Hunters engaged in killing, butchering and eating wild animals must have been exposed to all kinds of new zoonotic infections, sometimes with severe or fatal consequences. The rabies virus, for example, generally completes its life cycle in wild animals such as foxes, wolfs and bats, but if hunters were bitten by a rabid animal then the resulting encephalitis would be fatal. Butchering wild game was equally hazardous as the bacteria that cause tetanus, botulism and gas gangrene, all universally fatal, live naturally in animals' intestines. Similarly, eating raw or poorly cooked meat leads to the risk of ingesting the eggs of parasites like the tapeworm. Indeed even today Inuit bands hunting in the Arctic circle, where they eat a highly carnivorous diet, have a particularly high burden of these parasitic worms. However, each of these zoonotic infections is only acquired directly from their natural host and does not spread between humans. So although they must have infected individuals within a community from time to time, perhaps even the most active hunters, they are unlikely to have had a profound effect on the life of the band as a whole. But microbes like the trypanosome, that primarily resides in an animal host and has a flying vector (the tsetse fly), could have posed a major threat to the health of hunter-gatherer bands in Africa.

Sleeping Sickness (trypanosomiasis)

Although sleeping sickness (human African trypanosomiasis) has probably been endemic in Africa for many centuries, and indeed is reputed to have killed Mansa Djata, the Sultan of Mali, around 1374,[11] the first detailed descriptions date from the early twentieth century when an epidemic of 'Negro lethargy' swept across Central Africa. The disease is still a major problem with some 30,000 new cases reported annually.

Sleeping sickness is caused by a trypanosome, a highly active protozoan that swims in the blood with the aid of an undulating membrane attached along one side and a lashing terminal flagellum. The name derives from the Greek *trupanon* meaning borer, and refers to its corkscrew-like appearance. The disease begins in a non-specific way with fever, headache, enlarged lymph glands, skin rash and joint pains. But when the microbe invades the brain it causes the lethargy, drowsiness and coma that give the disease its name. Sleeping sickness is always fatal without treatment. Most experts think that hunter-gatherers could not have survived long term in the tsetse fly belt of Central Africa, and that the problems caused by sleeping sickness may have been the impetus for the human migration out of Africa which preceded the colonization of Europe and Asia some 50,000–100,000 years ago.

The sleeping sickness trypanosome was first identified in 1902 by Everett Dutton from the Liverpool School of Tropical Medicine, UK. While working in Gambia, West Africa, he spotted the microbe in the blood of an Englishman with a fever that did not respond to malaria drugs, and called the disease 'trypanosome fever'. At the same time a serious epidemic of sleeping sickness broke out in Uganda, but no one related this to trypanosome fever, and as the epidemic worsened, the Royal Society of Great Britain sent a Sleeping Sickness Expedition to investigate. In the end team members could not agree on the cause, although a young Italian bacteriologist, Aldo Castellani, was sure that the bacterium *Streptococcus* was the culprit. However, members of the Royal Society's Malaria Committee were unconvinced, and the following year David Bruce and David Nabarro were dispatched to Uganda on the same mission. Bruce, a Scottish army surgeon, had made his name in 1894 by discovering that *nagana* (the Zulu name for low spirits), a wasting disease of cattle which was rife in Africa, was caused by a trypanosome, now called *Trypanosoma (T.) brucei (b.) brucei,* and transmitted by the tsetse fly (*Glossina spp.*). Bruce and Nabarro soon found trypanosomes in the blood and spinal fluid of sleeping sickness victims and proved that they caused the disease, and were transmitted by tsetse flies, by successfully replicating the same infectious cycle in monkeys. Meanwhile, Castellani was back in Africa and this time he also found trypanosomes, but whether this was an entirely independent observation or influenced by Bruce's finding is far from clear. Castellani, of course, maintained the former and claimed the discovery as his own. Bruce rightly guessed that the trypanosome he had found was the same as the one discovered by Dutton in West Africa, but it was named after Bruce, *T.b. gambiense,* and is now known to cause sleeping sickness throughout West and Central Africa. In 1910 a third parasite, *T.b. rodesiense,* which is also spread by the tsetse fly, was discovered in the region that is now Zambia; this parasite causes sleeping sickness in East Africa. (These names can be confusing, so to explain: *Trypanosoma* is a genus of protozoa which contains the species *T. brucei,* of which there are three subspecies: *T.b. brucei,* which causes nagana in cattle, *T.b. gambiense* and *T.b. rhodesiense,* which both cause sleeping sickness in humans).

Today the sleeping sickness belt lies across the centre of Africa, and although the area it covers expands and contracts over time due to animal movements, human migrations and climate change, the disease has never taken hold outside Africa. This tight geographical restriction is dictated by the tsetse fly vector which requires the unique hot and humid conditions of the African tropics for its breeding cycle. The flies (both male and female) generally feed entirely on animal blood, searching out their prey by smell. They will travel for 90 metres up an odour plume to reach their victim and once they find it they suck its blood. That their meal may contain trypanosomes the flies neither know nor care. They feel no ill effects from this extra burden, and they unwittingly allow the parasites to multiply in their gut, ending up in their salivary gland ready to be injected into the next victim. In ideal conditions tsetse flies swarm in dense clouds around game herds, so the insect is a highly efficient vector for the trypanosome with an R_o value (the average number of new infections generated by one case) reaching levels as high as 388 among African cattle. Unlike mosquitoes, tsetse flies do not need water for their breeding cycle. The female nurtures a single larva inside her body, feeding it until the final stages of metamorphosis when she deposits it on the ground and it immediately pupates. The mature fly emerges two weeks later eager for its first blood meal. This cycle, taking around four weeks to produce just one adult fly, must be completed regularly and efficiently to maintain a critical population density of tsetse flies.

The two subtypes of *Trypanosoma brucei* that cause human disease have distinct geographical locations on either side of the African Rift Valley, and have different disease manifestations. In West Africa *T.b. gambiense* causes a chronic disease that may take many years to kill the victim, whereas in East Africa *T.b. rhodesiense* induces an acute illness which kills in six months. Today the meeting point of the two disease types is in Uganda, where there is *T.b. gambiense* in the north-west and *T.b. rhodesiense* in the south-east.

Of the three members of the *T. brucei* family, *T. b. brucei* infects wild game animals and domestic cattle, but not humans, East African *T.b. rhodesiense* infects both wild game and humans, and West African *T.b. gambiense* infects mainly humans. But although all three types look alike, molecular typing shows that the East African

T.b. rhodesiense is so closely related to the animal *T.b. brucei* that it must have evolved from it; indeed, the difference in a single gene gives *T.b. rhodesiense* the ability to infect humans. In contrast, the West African *T.b. gambiense* is quite distinct from the other two; no animal reservoir has been found and its origin is presently unknown, although recently bush pigs have been proposed as a reservoir.

When a bloodsucking tsetse fly ingests *T.b. brucei* from a wild animal and injects it into a human the parasite is immediately killed by contact with human serum. And although the mechanism of this killing is not understood, a single mutation at some time in the past was enough to convert *T.b. brucei* into its serum-resistant subtype, *T.b. rhodesiense,* which infects humans today. These two subtypes of *T. brucei* can happily coexist in the blood of virtually all wild mammals on the East African plains, from lions to antelopes to hyenas, causing them no problems. But when a tsetse fly transfers these parasites to a human only the mutated *T.b. rhodesiense* can take hold. Knowing exactly when this mutation occurred would date the first human infections with *T. b. rhodesiense*, and tell us for certain whether hunter-gatherers suffered from its effects, but this vital information is presently lacking.

Most experts think that the main foci of *T.b rhodesiense* that we see today in East Africa have existed for millennia and that the parasite infected hunter-gatherer tribes in Africa 50,000 years ago. And since the African Rift Valley is both the site of *T.b. brucei* infection in wild game and the ancestral home of humans, the two could have met early in human evolution.[12] At the critical point some 1.8 million years ago, when hominids moved from the rainforest to the open plains of East Africa and came into contact with herds of large game for the first time, they must have been bitten by trypanosome-carrying tsetse flies. At first the serum sensitivity of *T.b. brucei* protected them from infection, but as our ancestors became skilled big-game hunters so their exposure to tsetse flies and the parasites must have increased, and at some point a chance mutation in *T.b. brucei* overcame its serum sensitivity, allowing the new *T.b. rhodesiense* to survive in humans.

T.b. rhodesiense has adapted to both animal and insect hosts, causing no disease in either. This life cycle is well-balanced, stable and ancient, but when humans intrude on the cycle through being bitten by a parasite-laden tsetse fly they are not so lucky. The parasite has not adapted to this 'occasional' host; hence the disease progresses rapidly and is inevitably fatal.

To the west of the Rift Valley the ancestral trypanosome must have trodden a very different evolutionary path, although the details are far from clear. When hominids were first exploring the open plains in the east, the west remained cloaked in rainforest, and the ancestor of *T.b. gambiense* was probably a parasite of forest-dwelling apes. When this parasite first infected hominids or early humans, like *T.b. rhodesiense* in the east, it must have caused devastating disease. But then a human-adapted form evolved and dispensed with its animal reservoir so that *T.b. gambiense* and its human host could co-evolve to mutual benefit. Consequently, over thousands of years the parasite lost virulence, and today's sleeping sickness in West Africa is the milder, chronic form of the disease.

Since tsetse flies stay close to wild game, in hunter-gatherer bands the big-game hunters would be at greatest risk from the trypanosome. So the strongest, fittest and the most able hunters in a band were the most likely to be suddenly struck down with sleeping sickness, first becoming lethargic, then comatose and soon dying of the disease. In a fifty-strong hunter-gatherer band with ten or so fit, healthy hunters the loss of one or two could perhaps be sustained, but in the sleeping sickness belt of Africa it is likely that all those in regular contact with infected animals would be attacked by the microbe, leaving the band without any skilled hunters. Then band members could only fall back on trapping small prey and foraging for fruit and vegetables. But with the added problems of supporting the sick and dying, most of whom would have young families to be cared for, and the constant mobility necessary for finding sufficient food, it seems likely that many affected bands would have died out. Given this scenario, the lethal triad of the trypanosome, the tsetse fly and the sleeping sickness may account for the very slow growth rate of hunter-gatherer bands, which is estimated at only 0.003–0.01 per cent per year,[13] and for the eventual diaspora of hunter-gatherer bands out of Africa.

In more temperate regions hunter–gatherers encountered few new killer microbes and the resident herds of large game provided a ready food supply. So our ancestors entered a relatively healthy period of history with a plentiful food supply and the means to hunt it. But, as we shall see in the next chapter, eventually much of the wild game on each continent with the exception of Africa became extinct, probably at the hand of human hunters. Indeed, the protection of a few wild game species in Africa may be thanks to the trypanosome which drove man out of its territory, thereby preserving the animals it silently infects. In the next chapter we consider the far-reaching consequences of the extinction of wild life to our ancestor's lifestyle, with the subsequent human burden of infectious diseases.

3

MICROBES JUMP SPECIES

By moving out of Africa to colonize Asia and Europe, hunter-gatherers moved away from its lethal microbes to a healthier life and their population grew. In Eurasia big game was plentiful and as hunters became more adept with spears and clubs it was easy picking. For a time bands were almost exclusively carnivorous, probably dining off a single kill for a week or more. But this easy life was eventually jeopardized by restriction of their hunting grounds and loss of their prey.

When the last Ice Age began to lose its grip around 20,000 BC, the weather became warmer and drier and the landscape changed accordingly. Bands saw their traditional hunting grounds progressively eroded as the plains of Africa and Asia slowly dried into deserts, while the spear-throwing and club-wielding skills they had developed for hunting on open ground were less effective in the forests that replaced the grassland in more temperate regions.

Around the same time as this climate change came the demise of many of the world's largest animal species. By 12,000 years ago over 200 species had become extinct, including giants like the woolly mammoth and rhinoceros, sabre-toothed tiger, mastodon, and giant bison and sloth, all of which had risen to prominence in the Eurasian and African landscape after dinosaurs died out some 65 million years earlier. Both global warming and epidemic microbes have been blamed for this mass extinction, but although these may have contributed to the disaster, there seems little doubt that human hunters were the main culprit since the timing of the extinctions on each continent coincided with their arrival. So game animals were severely depleted in Africa by 40,000 BC, and by 20,000 BC they were virtually extinct in Eurasia. But the rate of extinction of these large animals was most dramatic in the Americas.

During the last Ice Age the sea level was low enough for the Bering Straits to form a temporary land bridge linking Siberia with Alaska, and many Old World mega-species crossed into the Americas, thriving there until humans followed them around 12,000 BC. Travelling from the north, these bands of nomads with their increasingly sophisticated hunting skills and ever-growing population depleted local herds and moved on southwards in search of more. With no previous human contact, herds of large-bodied herbivores made easy prey and were probably the first to go. Once herbivores were in short supply the food chain for scavengers and carnivores collapsed, and within as short a period as 400 years an estimated 135 American animal species vanished.

This fast-diminishing source of food inevitably led to competition between bands and most were forced to adopt the omnivorous diet of their ancestors. Archaeological records from the time show that bands now hunted smaller animals such as rabbits and deer, gathered fruits, grains and shellfish, and for the first time used boats for fishing. But still it must have been difficult to find enough to feed a large band and in many places mass starvation caused a precipitous fall in the population.[1]

This period of deprivation heralded a complete change in lifestyle from nomadic hunter-gatherer to sedentary farmer. Although in retrospect this change seems dramatic, in fact it must have evolved in slow stages as a result of changing circumstances, and in the end became a matter of necessity. Farming was adopted at different times in different places and in most areas farmers and hunter-gatherers probably coexisted for a while. But eventually, either voluntarily, or by force, persuasion, invasion, or even extermination, almost all hunter-gatherer bands gave way to the more successful farming communities.

The domestication of plants and animals evolved independently in at least nine places around the world, while all others imported domesticated species from these centres[2] (Table 3.1). The Fertile Crescent, the region between the Tigris and Euphrates rivers mainly in present-day Iraq and Iran, is famed as the earliest site of domestication. The first farmers grew emmer wheat and herded goats and sheep here from around 8500 BC, and this highly successful new lifestyle spread rapidly to surrounding areas of Asia, North Africa and Europe, where additional plants and animals more suited to the local conditions were soon added to the imported species. The Chinese began growing rice and keeping pigs around 7500 BC, and a little later several other centres in Africa (Sahel, tropical West Africa and Ethiopia) and Papua New Guinea also 'invented' farming. Interestingly, the farming lifestyle took hold much later in the Americas, perhaps because there were fewer plant and animal species suitable for domestication. However, from around 3500 BC Mexican Indians were growing corn, beans and squash and farming turkeys, while the people of the South American Andes famously domesticated the potato. By 2500 BC crops were also being grown in the eastern US but there is no evidence of animal domestication.

In theory a sedentary farming lifestyle has much to recommend it. It provides a permanent home to shelter the young, old and

Table 3.1 Examples of species domesticated in different areas of the world.

Area	Domesticated Plants	Domesticated Animals	Earliest attested date of domestication
Independent origins of domestication			
1. Southwest Asia	wheat, pea, olive	sheep, goat	8500 B.C.
2. China	rice, millet	pig, silkworm	by 7500 B.C.
3. Mesoamerica	corn, beans, squash	turkey	by 3500 B.C.
4. Andes and Amazonia	potato, manioc	llama, guinea pig	by 3500 B.C.
5. Eastern United States	sunflower, goosefoot	none	2500 B.C.
? 6. Sahel	sorghum, African rice	guinea fowl	by 5000 B.C.
? 7. Tropical West Africa	African yams, oil palm	none	by 3000 B.C.
? 8. Ethiopia	coffee, teff	none	?
? 9. New Guinea	sugar cane, banana	none	7000 B.C.?
Local domestication following arrival of founder crops from elsewhere			
10. Western Europe	poppy, oat	none	6000–3500 B.C.
11. Indus Valley	sesame, eggplant	humped cattle	7000 B.C.
12. Egypt	sycamore fig, chufa	donkey, cat	6000 B.C.

Source: Jared Diamond, *Guns, Germs and Steel* (1997), Brockman, Inc.

sick; a ready food supply with the facilities to store the excess; reliable sources of animal and plant materials for making clothes, blankets, ropes and tools; animals to assist with transport and farm work and to provide warmth in shared dwellings. With all this in its favour you might expect the life to be less arduous than hunting and gathering, for the people to be healthier, to live longer and for the population to grow more rapidly. But at first this was not the case. The tasks of digging, planting, harvesting and herding turned out to be more rather than less demanding than hunting and gathering, and archaeological records show that early farmers were smaller, less well-nourished, had a heavier disease burden, and died younger than their hunter-gatherer ancestors.[3] But quite quickly key discoveries like the plough and the wheel made farmers' lives easier and food production more efficient. Then their nutritional state improved, and it was not long before the birth rate began to rise. The interval between children in a family dropped from the hunter-gatherer's average of four years to one to two years, and in the resulting population explosion farming settlements grew into towns, some eventually becoming large cities.

The world's first towns and cities evolved from farming communities in the early centres of domestication, such as Jericho in the Fertile Crescent, and it is no coincidence that it was there too that catastrophic epidemics of infectious diseases first struck. These 'plagues' appeared suddenly as if from nowhere, swept through the population killing indiscriminately, and then disappeared just as mysteriously. As towns grew bigger and more densely populated, so these ferocious epidemics became more frequent and varied until they threatened the community's survival. So what were these new diseases, and where did they come from?

The change from a nomadic to a sedentary farming lifestyle that marked a turning point in human history also heralded a new era for microbes. For the first time humans had radically and permanently changed the landscape, disrupting naturally balanced ecosystems by clearing forests and scrub areas for planting, and reducing biodiversity by growing crops and farming animals. Microbes that had hitherto either not come into contact with domesticated plant and animal species, or had not been able to leap between isolated groups of individuals, were now handed the opportunity to flourish and many seized it. Presented with a whole field of wheat or herd of animals, microbes underwent a population explosion in these new host species. Although we have little specific information about the epidemic microbes that infected the first domesticated plants and animals, they must have repeatedly caused famines, spelling deprivation and even starvation for the early farmers.

In contrast to animal and plant microbes, we have plenty of information about the new infectious diseases that hit our ancestors during this period, many of which still cause us problems today. The unprecedented microbe bonanza was triggered by aspects of everyday life in the early farming communities that were absent from hunter–gatherer settlements: the build up of refuse, a high population density, and close contact with domesticated animals.

Without the constant moves typical of the hunter–gatherer lifestyle, all kinds of waste material, including human and animal faeces, accumulated in and around the permanent dwellings that families now shared with their livestock. And, lacking any knowledge of the danger it posed, the debris piled up and nobody bothered to clear it away. So early farming communities became a hotbed of parasites. Coprolites (fossilized faeces) found among archaeological remains from early farming communities commonly contain the eggs of intestinal worms, either passed directly from one person to another by faecal oral contamination, such as roundworms and hookworms, or transmitted by way of intermediate hosts, such as pork or beef tapeworms.[4] These parasites must have spread rapidly in the filthy surroundings, generally picked up by children and then carried for life. But although they may sometimes have caused anaemia from intestinal bleeding, worms of this type are not generally life-threatening and would have had little effect on the population as a whole.

For microbes that had difficulty finding hosts to infect among the sparse hunter-gatherer bands, the change to fixed communities was a new opening. The population density in the first villages was 10–100 times that of hunter-gatherer settlements, so microbes such as *Mycobacterium tuberculosis* (the cause of tuberculosis) that can only travel short distances between hosts, and *Mycobacterium leprae* (the cause of leprosy), that is too delicate to survive for long outside the human body, now spread with ease, and skeletal remains show that both TB and leprosy were more common in early farmers than in their hunter-gatherer ancestors.[5] But the most dramatic and long-lasting change in the pattern of human microbes at this time was the direct consequence of the domestication of animals. For the first time people were living in close contact with herd animals—drinking their milk, butchering and eating their flesh, curing their skins, caring for their young and sick, and sharing their shelter. Many animal microbes took the opportunity to jump ship, finding a new niche in a virgin human population.

There is now no doubt that most of the microbes that cause the classic acute childhood infectious diseases, such as smallpox, measles, mumps, diphtheria, whooping cough and scarlet fever, were originally exclusively animal pathogens that at some time in the past crossed the species barrier to infect humans. Today they only infect humans but their DNA sequences contain the tell-tale signs of their past lives. Their closest relatives are among the microbes of domestic animals, and in some cases the molecular clock even pinpoints the timing of their transfer to the early farming era. Perhaps at that time they caused epidemics of relatively mild infections in young animals, just like their human counterparts today. If so it is easy to see how farmers, having no reason to be cautious or to attribute human disease to their livestock, could have picked them up while tending their sick animals.

So our acute childhood infectious diseases were emerging microbes around 5000 BC, equivalent to the emerging infections of the twenty-first century like HIV, West Nile fever and SARS. Whereas we know the cause and can intervene to abort the natural cycles, when an epidemic hit our farming ancestors they had no notion of what to do. Paralysed with fear and panic, they often unwittingly assisted the microbe by carrying it with them as they fled, sometimes making for the apparent safety of the nearest town where the microbe could spread easily among the crowds.

Microbes do not usually jump species overnight but need time to evolve efficient ways of infecting and spreading between their new hosts. At first microbes often have to be picked up directly from their natural hosts, but generally it is only a matter of time before they adapt to humans and succeed in spreading directly between them. Then they can cause epidemics, and since no one in the early farming communities had either immunity or genetic resistance to these new microbes, epidemics would strike the whole community, causing severe disease, often with a high mortality. But the difference between our modern-day emerging infections and epidemics among early farmers is the number of susceptible human hosts that microbes had at their disposal. So whereas international airliners ferried SARS round the world, potentially exposing millions to the threat before it was even recognized, once an epidemic had infected everyone in a small isolated early farming community it had nowhere else to go and would die out. This abortive foray into the human population must have happened many times, perhaps over hundreds of years, before zoonotic microbes finally managed to establish their infectious cycles in humans independently of their natural animal hosts. Critical to this was the size of the population accessible to the microbe, since each one requires a minimum number of susceptible people within reach to establish its never-ending chain of infection. Most is known about measles epidemics in this regard, not least because when the virus was earmarked for global eradication it was essential to determine the size of the (non-vaccinated) population below which it could not survive.

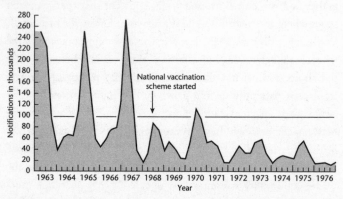

Figure 3.1 Notification of measles cases in UK from 1963 to 1976
Source: Public Health Laboratory Service, Community Disease Surveillance Centre.

Measles

The measles virus spreads between us with remarkable ease, and is proving a stubborn opponent in the fight to eradicate it. It caused huge epidemics among children worldwide until the first vaccine curtailed its spread in the 1960s (Figure 3.1). But the virus still causes serious outbreaks where it can get a foothold in pockets of unvaccinated people, and although the overall mortality from measles is less than 1 per cent, it can reach 40 per cent among malnourished children in developing countries. Today the virus kills around 350,000 children each year.

The measles virus colonizes the nasal passages of its victims, establishing a site of infection in the upper respiratory tract. From here it is shed in airborne droplets for several days before the typical measles rash appears. It has an R_0 value of 15, reflecting its high infectivity rate, spreading successfully to around 90 per cent of case contacts. All this is the hallmark of a virus whose life cycle has been highly tuned to its host by natural selection, but when it first encountered our ancestors it must have been a very different beast. The measles virus has many relatives among members of the morbillivirus family which infect a wide range of mammals, but it is most closely related to rinderpest virus of cattle, and to a lesser extent canine distemper virus. All three viruses probably arose from a common ancestor many centuries ago, and by using the molecular clock scientists pinpoint the divergence of rinderpest and measles viruses to around 2,000 years ago. This date rather satisfyingly suggests that the virus jumped from cattle to humans during the early farming period. Puzzlingly, though, the same type of analysis dates the most recent common ancestor of present-day measles viruses to just 100–200 years ago. The only possible explanation for this is that a strain with superior transmission powers spread round the world, at that time replacing all previous strains.[6]

Before a vaccine was available the rinderpest virus caused huge epidemics of 'cattle plague' that swept across Europe, Asia and Africa, attacking both domestic and wild herds and killing almost all of those it infected. In the 1890s it wiped out 80–90 per cent of the cattle in South Africa, but fortunately today an eradication campaign has pushed it to the brink of extinction. Like measles, the virus starts by infecting the upper respiratory tract, but then it attacks the gut, causing catastrophic diarrhoea and death from dehydration. Since the rinderpest virus spreads through contact with infected secretions and excretions, it is obviously aided by the herding instinct of wild cattle and the overcrowded conditions of some domestic herds. Although today many domestic breeds have some resistance to the virus, this must have been among the most devastating of cattle plagues for early farmers to cope with.

The ancestor of modern-day rinderpest and measles viruses probably jumped to humans on many occasions, sparking epidemics of a much more severe disease than we know today (indeed it was only differentiated from smallpox in the tenth century AD). But these first epidemics were probably restricted to the local inhabitants of villages and small towns, and once everyone was either dead or immune the infection would die out until another virus made the transfer from infected cattle.

To some extent this early pattern of infection is re-created by more recent measles epidemics in remote island communities. In 1846 one such epidemic hit the Faroes, a cluster of small Danish islands in the North Atlantic close to the Arctic Circle between Norway and Iceland. A carpenter developed measles shortly after arriving by boat from Denmark—the first introduction of the virus for sixty-five years. Over the next six months it infected

6,000 of the 7,782 islanders, only sparing those over the age of sixty-five who had suffered measles in the 1781 epidemic. But once the virus had run through the whole susceptible population it died out, not reappearing until the next time it was introduced from the outside world. Scientists studying this and similar epidemics in other island communities found the same epidemic pattern on islands with small populations, such as Iceland, Greenland and Fiji, with the chain of infection broken after each epidemic had run its course. Only on islands with larger populations like Hawaii could the measles virus maintain its infectious cycle and circulate continuously. Using this information, researchers calculate that the minimum number of people required to maintain the virus permanently in a city environment is around 500,000.[7] A similar figure probably applies to most other airborne acute infectious disease microbes, so when and where could these so-called 'crowd diseases' have first taken hold in the human population?

The world's oldest civilization arose in Mesopotamia (the Roman name for the Fertile Crescent area), where the first farming villages evolved from hunter-gatherer settlements on the fertile Sinjar Plain near modern-day Baghdad. As the population grew, villages expanded into towns and cities at the centre of trade, industry and government, probably reaching a population of 500,000 inhabitants sometime around 5,000 years ago. This date fits comfortably with the approximate date for the divergence of measles from the rinderpest virus, so this and many other zoonotic microbes probably cut their ties with their animal hosts and threw their lot in with humans around this time. And as these and other cities in the great civilizations of Egypt, Greece, India and China reached the critical size, so microbes must have taken hold in their inhabitants, causing devastating epidemics. Indeed, it is clear from ancient written records in the Egyptian medical papyri dating from around 1850 BC, Chinese medical texts from 1300 BC and the books of the Old Testament written between 1000 and 500 BC that epidemics were a major problem to these ancient civilizations, regularly killing a sizeable proportion of the population. In Exodus we read of the horrifying plagues of Egypt which, apart from unpleasantly large swarms of frogs, flies, locusts and lice, included a terrible pestilence in the form of 'boils breaking forth with blains'.[8] And in the first book of Samuel the poor Philistines were hit by 'emrods' of their secret parts (probably enlarged glands in the groin possibly caused by bubonic plague), which they interpreted as a divine punishment for stealing the Israelites' Ark of the Covenant, but when they returned it the plague spread to the Israelites as well.[9]

Leprosy is another frequently mentioned disease in the Old Testament and early Eastern texts, the first clear account being in the Indian *Charaka Samhit* written around 300–200 BC. *Mycobacterium leprae*, the bacterium causing leprosy, which spreads only poorly between humans, seems to have moved westwards along trade routes from India and China, causing a pandemic that reached its peak in the thirteenth and fourteenth centuries. Depending on the state of the sufferer's immunity *M. Leprae* may cause chronic skin lesions, destroy nerves, produce deformities, particularly of the face and hands, and invade and destroy internal organs. But in medieval times the term 'leprosy' was used to describe any chronic skin condition, and since 'lepers' were considered contagious, they became social outcasts, traditionally condemned to live in isolated colonies or to wear distinctive clothing and announce their presence by ringing a bell. We will pick up this subject again in Chapter 5 when considering yaws, another chronic skin infection that was originally included in the term 'leprosy'.

Despite the frequent and ferocious epidemics that hit the early civilizations of the Old World, over a longer timescale the population grew so that towns were continually expanding, growing ever more crowded, dirty and unhygienic as the years passed—an open invitation to pathogenic microbes. Looking back to these times, it is not possible to identify with any degree of confidence the microbes that caused the many epidemics described in ancient texts. This is partly because early descriptions often lack clinical details, and so diseases such as measles and smallpox where rash is a prominent feature are indistinguishable, but also because the diseases have very likely changed their character in the intervening years. Most would have begun by causing severe zoonotic infections; only once they were established as purely human pathogens could they co-evolve with their host, a process that takes around 150 years and generally results in milder disease.

Ancient Egypt

Much of the medical history of ancient Egypt is preserved in medical papyri and the embalmed bodies of Egyptian mummies, giving us a unique glimpse of the microbes that affected the people of one of the world's great early civilizations.

In Mesolithic times modern-day Egypt was populated by just a few thousand hunter-gatherers clustered in small isolated settlements along the Nile Valley. The change to farming began around 6000 BC when wheat, goats and sheep were imported from the Fertile Crescent. Conditions along the Tigris and Euphrates were very similar to those of the Nile Valley, and these new farming practices quickly became productive.

Egypt has very little rain, but before the Aswan Dam was built in 1902 the River Nile used to flood once a year, covering the valley for more than a mile on either side and depositing a rich sediment that kept the soil fertile. To preserve this valuable water, a system of irrigation farming evolved along the Nile as well as the Euphrates and Tigris Valleys, with a network of channels running between the fields to water the cereal crops. A similar system also developed in the Indus Valley in modern-day Pakistan, and on the Yellow River flood plain in China where it was used for watering rice paddies.

In Egypt this highly successful change in lifestyle produced the rapid population growth that heralded the ancient Egyptian civilization. The first large cities appeared around 2500 BC, at the same time as the famous sphinx and pyramids of Giza were built. Papyrus scrolls written by Egyptian physicians around this time describe common human ailments and their treatments. The Edwin Smith Surgical Papyrus, a 1700 BC copy of one thought to date from 3000 BC, details 'the pest of the year', an epidemic that struck Egypt annually.[10] Experts think that this may have been malaria arriving each year as the Nile flooded, providing an ideal breeding ground for mosquitoes. This suspicion is backed up by studies on 5,000-year-old mummies from Gebelein and Assyiout in the Luxor region of Upper Egypt, apparently a swampy area at the time. Just under half of the mummies show evidence of *P. falciparum* infection, and, interestingly, several also show signs suggestive of thalassaemia and sickle-cell anaemia, the inherited blood disorders that protect from malaria.[11] And since these genes take thousands of years to become common in malaria-exposed populations, this finding implies that the malaria parasite had been around in Egypt for a long time. Egyptian mummies also commonly contained intestinal parasites, and several skeletons show evidence of TB, but probably the most problematic infection at the time was schistosomiasis, a fatal disease caused by a waterborne microbe that exploits irrigation farming to aid its spread.

Schistosomiasis

This ancient parasitic disease is still a major public health problem today, infecting around 200 million people in over seventy countries, of which Egypt is a major focus. Descriptions that almost certainly refer to schistosomiasis, and remedies for treating it, are common in Egyptian papyri, particularly the Kahun papyrus dating from 1850 BC and the Papyrus Ebers of 1550 BC. The latter describes:

Another excellent remedy for those prepared for the belly.

Grind one *isu* and one *shames* finely and boil with honey. Should be eaten by a man in whose belly there are worms. The haematuria (bloody urine) produces them and they are not killed by any [other] remedy.[12]

Although there may be some debate about whether these descriptions really refer to schistosomiasis, the parasite's eggs found in the kidneys of two mummies dating from the XXth Dynasty spanning 1250–1000 BC leaves no doubt that the microbe infected the ancient Egyptians. And since schistosomes have also been uncovered in a well-preserved body buried in China around 200 BC, the microbe was obviously widespread in the ancient world.[13]

Schistosomiasis is caused by a parasitic blood fluke, the schistosome (derived from the Greek word *schistos*, meaning *'split'*, and *soma*, meaning 'body', and denoting the groove of the male in which he holds the female). The disease begins with an acute feverish illness as the parasite enters the body, but it is the long-term consequences of the infection, manifesting some two to ten years later, that are the most serious problem. The microbe's eggs cause inflammation, gouge out ulcers and induce scarring around the bladder or intestine. Depending on the site, the sufferer experiences chronic bloody diarrhoea, or haematuria,

Figure 3.2 Drawings from the Ebers papyrus possibly depicting haematuria caused by schistosomiasis

Source: Reprinted from *Infectious Disease Clinics of North America,* 18, A. A. F. Mahmoud, 'Schistosomiasis (bilharziasis): from antiquity to the present' 207–18, ©2004 with permission from Elsevier Inc.

with death from liver or kidney failure. The microbe can also predispose to bladder cancer, and even today schistosomiasis is among the leading causes of the tumour in endemic areas.

There are three main types of schistosomes, all with similar life cycles, but they use different species of water snail as intermediate hosts. Two of these (*S. mansoni* and *S. japonicum*) target the gut but the common type in Egypt (*S. haematobium*) has a particular predilection for the bladder, so haematuria has always been the most common and obvious symptom there. The microbe particularly affected young men, and indeed the Ebers papyrus contains a drawing of a penis supposedly discharging blood (Figure 3.2), a symptom so common that the Roman historian Herodotus is said to have referred to Egypt as 'the land where men menstruate'.[14]

The schistosome is sometimes called the Bilharzia worm after Theodore Bilharz, the German doctor who discovered it in human tissues, and its eggs in human excreta, while working at the Kasr el Ainy Hospital in Cairo in 1852. But he had no concept then of the parasite's complex existence, and there followed nearly sixty years of controversy before its mode of spread and life cycle, including its intermediate host, the water snail, were finally unravelled. For a while people thought that humans ingested the parasite, perhaps by

Figure 3.3 An illustration of a penile sheath worn in Ancient Egypt, circa XIX Dynasty 1350–1200BC

Source: Reprinted from *Infectious Disease Clinics of North America,* 18, A. A. F. Mahmoud, 'Schistosomiasis (bilharziasis): from antiquity to the present' 207–18, ©2004 with permission from Elsevier Inc.

eating spicy food or drinking contaminated water. But other routes also had their vogue—maybe it entered through the anus or urethra, possibly during masturbation. From early times the Egyptians realized that the disease was associated with water, and since the most prominent symptom was haematuria, they assumed it gained access through the urethra and devised protective penile sheaths to be worn while hunting in the marshes (Figure 3.3).

The schistosome attacked Napoleon's troops while they were stranded in Egypt in 1800, and British soldiers also suffered severely while fighting the Boer War in South Africa in 1899. With the outbreak of the First World War in 1914, and their troops heading for Egypt, the British government urgently needed the problem solved. In 1915 the British War Office sent parasitologist Lieutenant Colonel Robert Thomson Leiper to Cairo, with the express purpose of sorting out the parasites' transmission route.

He did not take long to come up with the answer: the ferocious little beasts burrow their way through the intact skin of people wading or swimming in contaminated water.

Flukes have separate male and female forms, and once they have got inside the body they make their way to the veins surrounding the intestine and bladder, where they mate. The female then deposits her eggs (hundreds to thousands a day—theoretically up to 600 billion in a lifetime!), which enter the bladder or gut lumen and return to the outside world in urine and faeces. In fresh water the eggs hatch and the new parasites then seek out their intermediate host, the water snail. They multiply inside this mollusc before once again being released into the water to complete their cycle by finding another human host to infect (Figure 3.4).

Water snails live in slow-flowing fresh water, and it is their habits that define the world distribution of the schistosomiasis. The irrigation channels along the Nile and in other areas of Africa and the Middle East, as well as the rice paddies of China and Japan, make ideal habitats, and with farmers wading barefoot in this contaminated water it is not surprising that the schistosome flourishes. But humans and water snails seem an unlikely pair of hosts for a parasite, so how did this combination evolve? Apparently, schistosome-like flukes have parasitized marine snails for at least 200 million years. The free-swimming larvae are clearly opportunistic, attempting to colonize any potential new hosts, leading to numerous host switches throughout history. At some stage they infected marine birds as a second host, and now use many mammal and fresh-water snail species, perhaps invading humans when irrigation farming brought them into close contact for the first time.[15]

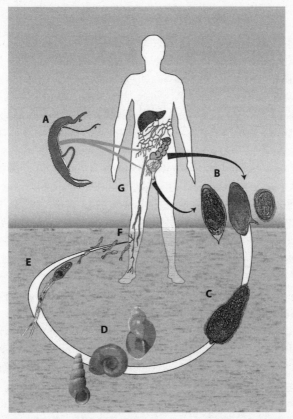

Figure 3.4 Schistosomiasis: transmission cycle of *Schistosoma mansoni*

Source: Reprinted from *The Lancet* 368, 23 September 2006, Bruno Gryseels and Pierre Wenseleers, 'Human Schistosomiasis', figure 1, page 1107, designed by the Institute of Tropical Medicine Antwerp, ©2006 with permission from Elsevier
Key: A: paired adult worms, B: eggs, C: ciliated miracidium, D: intermediate host snails, E: motile cercariae, F: cercariae penetrate human skin, G: cercariae migrate in blood vessels to the liver, mature, mate, and migrate to blood vessels around the bowel or bladder to lay eggs

Microbes Exploit Trade and War

As the population of the Old World grew and its towns and cities expanded each had its own local set of epidemic infectious diseases. Then, as trading networks opened up and reached previously isolated villages, so their microbes were shared and their infectious disease pools merged with those of the surrounding communities. And the more this happened the further microbes travelled until they were being carried across the Mediterranean and Indian Ocean by merchant ships, along with their cargos of corn, wine and olive oil. Similarly, caravans travelling the Silk Road to bring luxuries from China to the Middle East also brought less welcome visitors—their local microbes, spreading them to virgin populations. 'Plagues' became a constant feature of the times, and eventually a single infectious disease pool enveloped the whole of the Old World.

Like trade, warfare was a sure way for microbes to travel and spawn epidemics in virgin populations. As one after another, the great Egyptian, Assyrian, Persian, Greek and Roman Empires rose and fell, conflict, invasion and war were commonplace and inevitably sparked epidemics. With recruits drafted from widely differing backgrounds, housed in cramped and unsanitary army camps, stationed in overcrowded, war-torn towns, constantly on the move, and subject to stress, injury and malnutrition, it is no wonder that microbes decimated armies and spread from there to civilian populations. Time and again epidemics disrupted military operations and often determined the final outcome of wars. Although it is impossible to identify most of the microbes that beset the armies of the ancient world, three so-called plagues (although not necessarily caused by classic bubonic plague) of classical Greco-Roman times have been particularly well studied. These are the plague of Athens

74

in 430 BC, the Antonine plague in AD 166, and the Justinian plague in AD 542, each of which had far-reaching consequences.

The Plague of Athens

The legendary rivalry between the great cities of ancient Greece regularly erupted into the conflict that dogged the empire and ultimately led to its downfall. And although their separate armies had united to defeat the Persians when Xerxes threatened to invade in 480–479 BC, old jealousies soon resurfaced. By 431 BC Athens and Sparta were again at loggerheads in the Peloponnesian War, which lasted on and off for twenty-seven years. Athens with her mighty navy was pitted against the vastly superior and well-disciplined Spartan infantry, and Pericles, ruler of Athens, resolved not to confront the Spartans head on. Instead he decided to fortify the city and sit out the siege until the Spartans called for a truce. He enclosed Athens, including the port of Piraeus, within wooden walls, intending to use the port's access to the sea as the city's lifeline. When all agreed to this ingenious strategy to pre-serve the city by avoiding direct conflict, no one could have foreseen the terrible consequences. As the Spartans advanced, thousands of refugees from the surrounding countryside headed for the safety of Athens, so that the city with its mere 10,000 dwellings, and no route out except by sea, housed a seething mass of people—an ideal setting for a microbe to take hold. When the epidemic struck in 430 BC it was so virulent and widespread that it spelt defeat for the Athenians, thus contributing to the end of the golden age of Greek culture and their dominance of the ancient world.

This is the oldest epidemic ever recorded, documented by the contemporary historian, Thucydides. According to him, it began abruptly and rampaged for four years, killing around a quarter of the population, military and civilian alike. The Athenian army lost 4,400 foot soldiers and 300 cavalrymen, over a quarter of their front line. But most importantly Pericles himself was among the casualties, and his death not only deprived the Athenians of their great leader but also demoralized a people already ravaged by war and plague.

No physical evidence of the plague survives, so to decide on the specific nature of the culprit microbe experts must interpret Thucydides' text as best they can. He is clear that the disease affected all ages, killed in seven to nine days, describing it thus:

internally, the heat was so intense that the victims could not endure the laying-on of even the lightest wraps and linens; indeed nothing would suffice but that they must go naked, and a plunge into cold water would give the greatest relief. Many who were left unattended actually did this, jumping into wells, so unquenchable was the thirst which possessed them; but it was all the same, whether they drank much or little.[16]

Thucydides also lists the other symptoms (starting with sore eyes and throat, cough, sneezing, blisters, sores and foul breath; proceeding to retching, convulsions and a red rash; and finally diarrhoea, sleeplessness, gangrene and loss of memory and sight, and death from exhaustion), but his description does not tie up with any infectious disease we know today. Smallpox is favoured by most experts; measles or typhoid by others[17]—indeed it may have been a mixture of infectious diseases, all raging in the appalling conditions created by the siege.

Interestingly the Spartan army seemed to be spared the plague, so perhaps they were already immune to the microbe. It is even possible that they were the source of it, bringing along with their arms a disease which they were already familiar with but which was new and lethal to the Athenians.

The plague of Athens was devastating, not just because of the huge numbers it killed but by its effect on the living. Thucydides describes the change in personal attitudes thus:

'They saw how sudden was the change of fortune in the case both of those who were prosperous and suddenly died, and of those who before had nothing but, in a moment, were in possession of the property of others.'[18]

Incidentally, the Greek Empire was similarly left rudderless in 323 BC by the sudden death of the legendary Alexander the Great while at the height of his power. This extraordinary young man led his Macedonian troops against the Persians, conquering their great empire and eventually controlling an area stretching from Egypt to the Indus Valley. But then, shortly after sailing down the Euphrates from Babylon to inspect the marshlands near the Arabian border, he developed a fever, and in less than two weeks this healthy thirty-three-year-old leader was dead. There were no epidemics around at the time, so it seems reasonable to assume that Alexander's death was related to his marshland trip. It could have been malaria, but he suffered from continuous fever for eleven days which sounds more like typhoid, caused by a bacterium spread by faecal contamination of food or water. After his death the empire was split between his generals and a period of unrest ensued that contributed to the demise of his greater Greece.[19]

The Antonine Plague

This plague, the least well documented of the three, hit the Roman Empire at its height. From the magnificent city of Rome with more than a million inhabitants, Emperor Marcus Aurelius Antoninus ruled over an area stretching from Britain in the West through Europe to the Middle East and North Africa. The very extent of this multicultural, multinational empire, with its traders free to operate in any province and its armies constantly on the march, provided a network of highways for microbes which could hitch a ride with travellers and spark off epidemics wherever they went.

The source of the Antonine plague was probably the city of Seleucia on the River Tigris near modern-day Baghdad. Roman troops sent to quell an uprising sacked and plundered the city and then returned triumphant, spreading the plague along their route and bringing it home to Rome, where at its height it killed around 5,000 people a day. Eventually it spread over the whole empire and beyond to India and China, lasting several decades, and still raging when the Emperor Marcus Aurelius died in AD 180.

The renowned physician Galen of Pergamum relates in his *Methodus Medendi* that the disease struck indiscriminately at rich and poor, young and old alike, killing between a third and a half of those it infected. He describes a 'fever plague' causing intense heat and thirst, vomiting and diarrhoea, all very similar to the plague of Athens, but this time there is a clear description of a dry, black, ulcerated rash covering the whole body, which Galen attributes to 'a remnant of blood which had putrefied in the fever blisters'. This description, particularly of the rash, leaves little doubt that the plague was smallpox, possibly the first of its kind in Europe. But some experts still favour typhus as the cause, maintaining that in its early stages it cannot be distinguished from smallpox.[20]

The Romans believed the plague was divine punishment for their deeds in the city of Seleucia, where soldiers sacked the temple of Apollo and opened an ancient sealed tomb. As the contemporary historian Ammianus Marcellinus wrote, 'when the Roman soldiers opened it up, pestilence issued forth, bringing contagion and death all over the empire from the borders of Iran to the Rhine river and Gaul'.[21]

The Antonine plague caused such a calamitous drop in population that the Roman Empire, critically reliant on manpower for its every function, began to founder. Towns and fields stood empty, the army was depleted, trade and commerce stagnated, and the people were confused and demoralized. This marked the onset of a decline lasting for the next 100 years, with constant invasions, wars and plagues. In AD 266 the Emperor Valerian was taken prisoner by the Iranians and the extremities of the empire to the East and West were lost. These struggles continued until Constantine the Great moved the capital of the empire to Byzantium, renaming it Constantinople, and in 396 the empire split into its Latin Western and Greek Eastern halves.

The Justinian Plague

After the centuries of strife, the Emperor Justinian managed briefly to reconquer North Africa, Italy and Spain, and to reunite the Roman Empire in the sixth century AD, but when plague struck Constantinople it heralded a series of epidemic cycles that lasted for two centuries. It marked the end of the empire, the relative isolation of Europe in the West and the expansion of Islam in the East. In Constantinople the Justinian plague lasted for a year, killing a quarter of the population—10,000 people a day at its height. Constantine himself survived an attack, but as the epidemic surged both East and West throughout the empire it left few in its wake to perform essential tasks and he was powerless to protect the newly united territories. The overall death toll in the two empires is estimated at 100 million.

The Byzantium chronicler Procopius of Caesarea states that the plague began in Pelusium in Egypt, spreading from there to Alexandria and on via Palastine to Constantinople. He described the symptoms in such detail that most experts are in no doubt that this was bubonic plague with its characteristic hugely swollen glands (buboes):

The fever, from morning to night, was so slight that neither the patients nor the physician feared danger, and no one believed that he would die. But in many even on the first day, in others on the day following, in others again not until later, a bubo appeared both in the inguinal regions and under the armpits; in some behind the ears, and in any part of the body whatsoever.

To this point, the disease was the same in everyone, but in the later stages there were individual differences. Some went into a coma; others violent delirium. If they neither fell asleep nor became delirious, the swelling gangrened and these died of excess pain. It was not contagious to touch, since no doctor or private individual fell ill from the sick or dead; for many who nursed or buried, remained alive in their service, contrary to all expectations.

Some died at once; others after many days; and the bodies of some broke out in black blisters the size of a lentil. These did not live after one day, but died at once; and many were quickly killed by a vomiting of blood which attacked them. Physicians could not tell which cases were light and which severe, and no remedies availed.[22]

This was perhaps the first epidemic of bubonic plague in Europe; it raged for 200 years, mysteriously disappeared for 600 years, and then, as we shall see in the next chapter, reappeared in the guise of the Black Death.

We do not know for certain the cause of any of these epidemics, and earlier writers were confident that we never should. But now that new molecular probes can detect the faintest fingerprints of microbes, finding the answer is a real possibility. Plague victims are being excavated from mass burial sites, so it may just be a matter of time before the microbes that caused these devastating epidemics and pandemics are revealed.

Some time around AD 1200 these terrible, unpredictable epidemics gave way to a cyclical pattern, with microbes visiting a community and causing an epidemic only when there were enough susceptible people to sustain their chains of infection. And since recovery from these acute infectious diseases gives lifelong protection, only children born since the last epidemic were susceptible to the next. So the pattern of childhood epidemics, like that of measles illustrated in Figure 3.1, was established and, as each cycle weeded out the most susceptible children, over generations resistance built up in the population and the infections became less severe. But as we shall see in the next chapter, the acute infectious diseases remained a threat, especially to children in towns and cities where increasing crowding, poverty and unsanitary conditions encouraged their spread.

4

CROWDS, FILTH AND POVERTY

In medieval Europe virtually everyone was infested with blood-sucking parasites and their homes were colonized by equally infested mice and rats. Most people in farming villages lived along with their livestock in tiny, single-storey, thatched hovels which were dark, cramped and airless. As the population expanded and small communities grew into towns, the situation got steadily worse. With no facilities for waste disposal everything was thrown out into the narrow lanes that ran between the dwellings so that these dark, dank conduits became quagmires of mud, human and animal excreta and garbage, most of which ended up in the rivers that served as the water supply. In these unhygienic surroundings it is no wonder that microbes flourished. Almost all their transmission routes were facilitated by the appalling conditions: poor ventilation and overcrowding made life easy for airborne microbes; non-existent sanitation meant that gut pathogens had easy access to food and water; and lack of personal hygiene allowed vectors like fleas and lice to prosper.

Not surprisingly, medieval towns and cities were extremely unhealthy places to live; their inhabitants had a lower life expectancy than their country cousins, and it was not until the beginning of the twentieth century that European cities could sustain their own population.[1] But in medieval Europe the vast majority of people lived and worked on country estates where a distinct class system operated, with peasants tied to the manor, working land which belonged to the Lord of the manor in return for services rendered. Towns and cities relied on these surrounding estates not only for imports of consumables like grain, meat and wood, but also for manpower, attracting country youths to the urban life with the prospect of better wages. All too often this was a one-way ticket to death from one of the acute infections endemic in towns, which they had hitherto escaped in the healthier countryside.

Despite the unhealthy environment in its cities, Europe experienced an unprecedented population explosion during the eleventh and twelfth centuries, and by the middle of the thirteenth century people were outgrowing natural resources. Good agricultural land was in short supply and severe under-employment inevitably led to poverty. By the fourteenth century, with the Little Ice Age well under way, the drop in temperature caused crop yields to fall and regular famines ensued.

Throughout this period of population growth more people travelled further than ever before, be it for the purposes of trade, war or pilgrimage, taking their microbes with them and broadcasting them like seeds along the way. Merchant ships crossed the Mediterranean with ever-increasing frequency, and Europeans set up trading centres as far afield as Africa, India and the Far East. At the same time local skirmishes between European cities as well as international wars meant that troops were always on the move, while the Crusades of the eleventh, twelfth and thirteenth centuries saw large Christian armies heading East across Europe for Jerusalem to challenge the Saracens. More often than not crusaders were turned back by epidemic diseases like dysentery, typhoid and smallpox that ravaged their troops and sparked epidemics in their home countries.

Meanwhile, the vast Eastern Mongol Empire established under Genghis Khan eventually incorporated the whole of modern-day China and most of Russia, and stretched through central Asia to Iran and Iraq in the West. Its capital, Karakorum, was the hub of a trading network linking all the large and flourishing cities of the empire and beyond to Europe. The old Silk Road between China and Syria was re-established and used by literally thousands of traders, soldiers, post riders and ambassadors travelling from oasis to oasis and covering vast distances. When the Venetian Marco Polo (1254–1324), set out on his epic journey in 1271, travelling the Silk Road through Armenia, Persia and Afghanistan, over the Pamirs and through the Gobi Desert to Beijing; a distance of 5,600 miles and taking three and a half years to complete, he certainly never spared a thought for the microbes he might pick up or pass on along the way. But there is little doubt that the opening up of this and other international trading routes, with the inevitable exchange of people, their animals, food stuffs and materials, facilitated the spread of microbes.

By medieval times most of the acute infectious diseases were universal in the Old World and had settled into distinct cycles of epidemics, mainly affecting young children. Many had already co-evolved with humans to become less virulent, but plague and smallpox remained fearsome killers over several centuries, standing out as the most lethal of the times. Throughout history each has probably killed more people than all the other infectious diseases added together, between them holding the human population in check for several centuries.

Bubonic Plague

The world has seen three pandemics of bubonic plague in the past 2,000 years: the first was the Justinian plague in 542 (see Chapter 3), the second the Black Death, and the third, which began in China in the 1860s, is still ongoing today.

The legendary Black Death was only so called hundreds of years after the event and now no one knows why. Some say the name refers to the gangrene that sets in, turning victims' fingers and toes black just before death, but others are sure that it derives from a mistranslation of the Latin *atra mors* since *atra* can mean either 'terrible' or 'black'. The Black Death raged between 1346 and 1353, eventually spreading through the whole of Europe, Asia and North Africa. The disease hit almost every city, town, village and hamlet, wiping out nearly half the inhabitants and causing the most dramatic fall in population ever recorded. The pandemic spread to Europe from the East, and can be traced to the Golden Horde (the Eastern section of the Mongol Empire located in present-day Russia), where a ferocious epidemic was raging, reaching the port of Caffa on the Black Sea in 1347. At the time this major Genoese trading centre was under siege by the Mongols, but when the epidemic broke out among their troops they were forced to withdraw. Some say their parting shot was to catapult their dead into the city, but whatever the truth of it, the epidemic soon spread to the besieged inhabitants of Caffa. Many Italian residents hastily set sail for home, and twelve of their galleys made for Messina on the East coast of Sicily. No sooner had they put into port than plague broke out—a disease as horrifying as it was devastating. In the words of Michael of Piazza, a Franciscan friar who wrote a first-hand account:

The 'burn blisters' appeared, and boils developed in different parts of the body: in the sexual organs, in others on the thighs, or on the arm, and in others on the neck. At first these were of the size of a hazelnut and the patient was seized by violent shivering fits, which soon rendered him so weak that he could no longer stand upright, but was forced to lie on his bed, consumed by a violent fever and over-come by great tribulation. Soon the boils grew to the size of a walnut, then to that of a hen's egg or a goose's egg, and they were exceedingly painful, and irritated the body, causing it to vomit blood by vitiating the juices. The blood rose from the affected lungs to the throat, producing a putrefying and ultimately decomposing effect on the whole body. The sickness lasted three days, and on the fourth, at the latest, the patient succumbed.[2]

Nothing like it had ever been seen in Sicily before, and as the corpses piled up the terrified inhabitants of Messina panicked and fled. The ships' crews were blamed for the disaster and expelled from the island. They, apparently still unaffected, joined other galleys from Caffa sailing to Genoa and Venice, carrying the deadly microbe with them.

From these Italian cities the microbe was ferried to most Medi-terranean ports, and then taken inland along well-trodden trading routes, or carried, often unwittingly as in Messina, by people fleeing from plagued towns and cities. The pandemic engulfed Europe like a tidal wave, reaching France, Spain and the Mediterranean islands in 1347–1348, and covering the whole continent in just three years (Figure 4.1). Virtually every town, village and hamlet was caught in its deadly grasp, and in each the plague raged for around eight months, only remitting when almost everyone had been infected and was either dead or recovered.

The microbe reached British shores in the summer of 1348, probably sneaking in from France through the port of Melcombe Regis on the south coast (now part of Weymouth). The pandemic

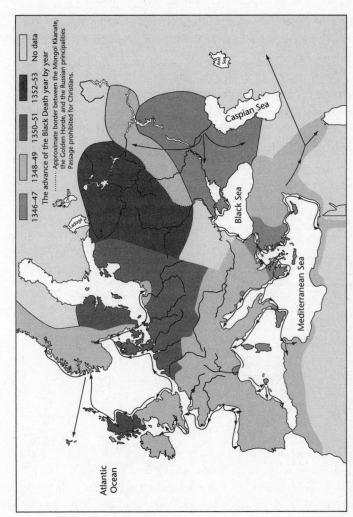

Figure 4.1 Map showing the advance of the Black Death

hit London later that year, killing about 20,000–30,000 of its 60,000–70,000 inhabitants. Then, travelling northwards at a rate of 1–1.5 km per day, it covered the length of England (500 km) in around 500 days.[3]

At the time of the Black Death the microbe had not visited Europe for a thousand years and so it struck indiscriminately and ferociously, killing old and young in town and country. People seemed in no doubt that the plague spread directly from one person to another, and they naturally took precautions to prevent it. Quarantine (from the Italian *quaranta giorni*, meaning 'forty days') was first introduced in Italy and later became widely practised throughout Europe. Plague-stricken towns were sealed off and ships arriving from affected ports were isolated for forty days to prevent any person or merchandise coming ashore until all danger of infection was past. Homes of victims were marked with a red cross, and the inhabitants either isolated or removed to pest houses—hospitals appropriately described as 'simply waiting rooms for death'.[4] Of course the rich could escape the infection, as royal courts often did, by moving to the country for the duration of the plague in the city, and in contrast to smallpox which devastated European royal families, this precaution seemed to have some effect against the plague since only one European monarch died of it. This was King Alfonso XI of Spain, who was involved in a battle with the Arabs in Gibraltar when the plague struck and refused to leave his troops.[5] On the other hand, Pope Clement VI survived by isolating himself from the outside world and, on the advice of his doctor, sitting alone in his study between two huge open fires while the epidemic raged in Avignon.[6]

Although the plague in Europe finally abated in 1353, it resurfaced unpredictably for the next 300 years, causing terrifying epidemics that were often centred on towns while sparing the surrounding countryside, but still with the same high death toll among its victims. France and Italy, having the densest populations, were worst hit with the disease raging almost continuously in one area or another. Britain was less affected by these ravages since the microbe could not be sustained in its sparse island population and each epidemic had to be introduced from the continent.

The Renaissance plague (also known as the Great Plague of London) in 1665–6, was the microbe's final outburst, after which it disappeared completely from Northern Europe. The famous diarist Samuel Pepys remained in London throughout the Great Plague, watching as the death toll rose alarmingly through the summer of 1665. In August he wrote in his diary:

In the city died this week 7,496, and of them 6,102 of the plague. But it is feared that the true number of the dead this week is near 10,000; partly from the poor who cannot be taken notice of, through the greatness of the number, and partly from the Quakers and others that will not have any bells rung for them.[7]

The capital city was still no bigger than it had been at the time of the Black Death 300 years before, but its population had increased tenfold so it was now even dirtier and more crowded. As the microbe spread relentlessly the rich left for healthier surroundings, abandoning the poor to their fate; some survived the plague only to die of starvation. For a while anarchy reigned; many took on the tasks of nursing the sick and guarding quarantined houses with the express purpose of bribing, robbing or even murdering their helpless plague-stricken victims. The dead were thrown into the street, where they lay until collected by the death cart to be flung into hastily dug burial pits. Quackery abounded, but of course there were many doctors, apothecaries and nurses who did what they could until all too often they succumbed to the

microbe. Not surprisingly, despite protecting themselves with lea-
ther costumes, spice-containing beaked masks and wands of in-
cense, surgeons responsible for blood-letting and knifing buboes
had the highest death rate (see cover picture of a plague doctor).

As often happened with the plague, the epidemic seemed to
abate during the winter but it reappeared again in the spring of
1666 just as the rich thought it safe to return to the city. And
although the worst was over, another 2,000 died in London before
the epidemic finally fizzled out towards the end of 1666. The
official death toll was 68,595, giving a death rate of 15 per cent,
but as the rich had left, and, as Pepys pointed out, many deaths
went unreported, the true number was probably significantly
higher.

Of course the people of medieval Europe had no inkling that a tiny
microbe might be responsible for their terrible suffering, and they
variously blamed the rare conjunction of the planets, miasmic vapours, or the wrath of God. The true cause of the plague was only
uncovered at the beginning of the twentieth century after an epidemic hit Hong Kong, starting the third pandemic. Plague that had
been smouldering unnoticed by the outside world in the Yunnan
province of China since the middle of the nineteenth century
reached Canton in 1894, wiping out 40 per cent of its 100,000
inhabitants. But it was only when the microbe invaded the British
port of Hong Kong and threatened commercial interests worldwide
that things began to happen. This was the golden age of bacteriology
when Louis Pasteur and Robert Koch were busy identifying microbes and propounding the germ theory of infectious disease (see
Chapter 8), so when the cry for help came, two experienced bac-
teriologists took up the challenge. Alexander Yersin was a shy, young
Swiss microbiologist, who had trained under Pasteur in Paris and

travelled widely in the Far East, while Shibasaburo Kitasato was an imperious Japanese professor already famous for his groundbreaking work on tuberculosis and tetanus with Koch in Berlin. By the time Yersin arrived in Hong Kong in June 1894, Kitasato was already installed in his own laboratory with five assistants working for him. He claimed to have found the 'plague bacillus' in the blood of victims and was preparing to publish his findings in the medical journal, *The Lancet*.[8] But Yersin, who was given no help and had to build himself a straw hut to house his laboratory, could find no bacteria in the blood of plague cases. He was keen to search for the culprit microbe in buboes but he was denied access to the bodies of plague victims, which were all reserved for Kitasato. Rivalry between the two was intense and in the end Yersin had to bribe mortuary attendants for access to the material he needed. But once he had it he turned up trumps. He discovered the microbe now called *Yersinia (Y.) pestis* in his honour, and published his findings shortly after Kitasato.[9] This sparked off a controversy that continued for decades; most initially believed that the eminent bacteriologist Kitasato had to be right and only slowly came to realize that his work was flawed. Yersin had found the true plague bacterium but sadly the bacterium was not named after him until 1970, long after his death.

At the time of the Hong Kong epidemic most doctors, including Kitasato, thought that the plague microbe hid in soil, jumping directly from this reservoir to infected humans. Only Yersin heeded reports of dead rats littering plague-infested cities and took the trouble to investigate. He suggested that rats were the main vehicle of the infection but it was left to his successor from Pasteur's laboratory, the equally reclusive Paul-Louis Simond, to figure out the microbe's life cycle, including its reliance on rat fleas to carry it to humans, when working in India three years later.

While the controversy over the identity of the plague microbe raged, the microbe itself silently boarded ship in Hong Kong for destinations in India, the Middle East, Africa, Europe, Russia, Japan, Australia, the Americas, Indonesia and Madagascar. In most places the outbreaks it spawned were soon under control, but the microbe gained a foothold in the US which has never been eliminated, and in India it caused a devastating epidemic killing more than 10 million people. In 1905 The Indian Plague Research Commission was set up to unravel the epidemiology of *Y. pestis* and devise means to stop its rampant spread. This body repeated Simond's work, confirming that the microbe primarily infects rodents but uses rat fleas to ferry itself to humans. With that, rat extermination became the mainstay of plague control in infested cities and remained so until the advent of antibiotics converted this fearsome killer into a treatable disease.

Y. pestis naturally infects burrowing rodents such as gerbils, marmots and ground squirrels, and is spread among them by their blood-sucking fleas. Rodents show varying degrees of susceptibility to the microbe and in colonies of resistant animals it can circulate continuously without causing any ill effect. There are several such colonies, called plague foci, in the world today (Figure 4.2), which act as reservoirs of the microbe. Most of these, including the foci in California, South Africa and Argentina, were only established during the third pandemic when microbes from Hong Kong found new hosts among local burrowing rodents. In the late 1890s *Y. pestis* arrived in San Francisco, probably carried there by stowaway rats on an Asian merchant ship, and causing a small epidemic among Chinese immigrants. Almost immediately ground squirrels picked up the microbe and it now circulates in the US, where there are over fifty rodent species to act as potential carriers.

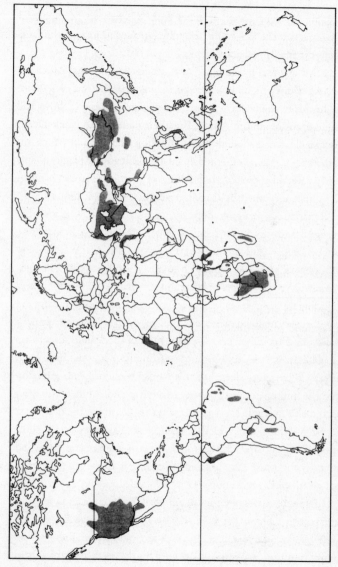

Figure 4.2 Map of present day plague foci worldwide

Source: Reprinted from *Medical Microbiology*, 16th edn, 2002, David Greenwood, Richard C. B. Slack, John F. Peutherer. Fig. 35.1, page 332. ©2007, with permission from Elsevier Ltd.

This huge plague focus stretches from Canada to Mexico and halfway across the US, so the microbe is poised to jump to humans whenever the opportunity arises.

The plague foci in the Himalayas, Eurasia and Central Africa are considered ancient, and most experts agree that the first pandemic, the Justinian plague of AD 542 (see Chapter 4), arose from the plague focus in central Africa. But the origin of the Black Death is not so clear-cut; some think it came from the focus in Africa but most believe that it emerged from the Himalayan plague focus, perhaps aided by the invading Mongols who were in the habit of killing (possibly microbe-carrying) marmots for their fur.

Remarkably, recent genetic studies show that Y. pestis only diverged from its closest relative, Y. pseudotuberculosis, some time between 1,500 and 20,000 years ago, so it is a relatively new pathogen to both rodents and humans.[10] And since Y. pseudotuberculosis is a gut microbe that infects many mammal species including rats, and spreads in food and water, it must have undergone several major genetic changes to evolve into the virulent flea-borne Y. pestis we know today. Y. pseudotuberculosis sometimes circulates in the blood of rodents so from there it could have been picked up by biting fleas, but this would be a dead-end infection until the microbe could not only survive in and colonize the flea's gut but also multiply and spread from the bite site in the flea's next victim. Most experts agree that even 20,000 years is way too short a period for all these changes to evolve by random mutation, so the microbe must have picked up plasmids from other bacteria that completely revolutionized its life cycle.

Human plague pandemics begin with Y. pestis escaping from its reservoir to infect other wild rodents, and this is usually precipitated by favourable climatic conditions and an abundant food supply resulting in overpopulation. This drives the inhabitants of the plague focus to forage over a larger area, where they are more likely to make contact with (and share fleas with) other species. Some of these species are highly susceptible to *Y. pestis;* rats in particular die within days of infection, and it is these animals that form the critical link to human epidemics.

In most of Europe today the brown (sewer) rat *(Rattus norvegicus)*, a hardy creature whose origins are in Russia, is the commonest type, but since it only arrived in Britain some time after the last plague outbreak it could not have been responsible for spreading the microbe at the time of the Black Death. This dubious accolade goes to the black rat (*Rattus rattus*) or house rat, which, unlike the brown rat, is not all that hardy. Its ancestral home is in India in the foothills of the Himalayas, but long ago it spread throughout the tropics. Black rats were well established in North Africa by the beginning of the Christian era. Then, as international trade routes opened up, the rats followed along. They stowed away on ships, crossed the Mediterranean and colonized ports along all the main shipping routes. Then they spread throughout Europe, accompanying caravans of traders on their journeys from town to town, reaching Britain some time in the Middle Ages. They found a niche in colder regions by cohabiting with humans, making nests in the thatched roofs of houses, in granaries and barns, and, being territorial animals, each rural household generally supported one distinct colony. But in towns and cities black rat colonies knew no boundaries—the crowded, squalid dwellings were simply overrun with them. So by medieval times the black rat was well placed to act as an intermediary host for *Y. pestis*, and their fleas (each rat has an average of three) as the vehicle that carried it to man.

Once *Y. pestis* gets a foothold in a black rat colony, and is passed between the residents by their fleas, rats die rapidly. Fleas carrying *Y. pestis* desert the dying for the living, so after ten to fourteen days the microbe has virtually wiped out the whole colony. By then the place is literally swarming with desperately hungry fleas, and although human blood is very much second best for rat fleas, at this stage they are ready to try anything, and one bite is enough to deliver a potentially lethal injection to humans.

There are 2,500 different types of flea and not all can transmit *Y. pestis,* but the rat flea (*Xenopsylla cheopis*) seems especially designed for the task. Its stomach has an entry valve that allows it to become fully distended with blood without regurgitating the contents the next time the insect feeds. But after a blood meal containing *Y. pestis* the bacteria multiply to form a ball of microbes mixed with blood in the flea's stomach that inactivates the valve. So the next time the flea tries to feed it regurgitates its stomach contents, now containing up to 25,000 bacteria, into its new victim. And not content with that, the frustrated flea, still hungry, goes on desperately biting and spreading the bacteria until it literally starves to death.

So plague victims are not directly infectious to others, and fleas cannot transmit the disease between them because in human blood *Y. pestis* does not reach the high concentrations required to block a flea's stomach. But rat fleas, unlike the human variety, actually cling on to their host and travel with them. So if a visitor to a plague victim's home unwittingly picks up a *Y. pestis*-carrying rat flea it will hitch a ride to their home, not only infecting the carrier but also their local rat population, setting the whole cycle in motion and producing a new crop of human disease some twenty-four days later. Fleas can also survive for some time without their rat or human hosts, particularly in cool moist conditions, so *Y. pestis* can remain alive on long sea voyages even if it has killed the resident rat population before the ship arrives in port.

Fleas usually bite exposed areas of skin like the face, arm or leg, and once injected *Y. pestis* travels to, and multiplies in, the local lymph glands in the neck, armpit or groin respectively. And although the immune system is alerted, this microbe has a veritable armament of devices ready to foil the attack. It grows happily inside macrophages when they engulf and try to kill it, and it tricks the body into overproducing suppressive cytokines that knock out key immune cells. This strategy gains time, and the microbe can multiply to enormous numbers, causing the glands to swell into the characteristic buboes—huge, exquisitely painful abscesses. If the immune attack succeeds in restricting the microbe to the gland then the victim stands a chance of surviving, particularly if the buboes rupture and discharge their stinking pus. But all too often *Y. pestis* is one step ahead, spilling out into the bloodstream, attacking blood vessels and causing bleeding into vital organs. Skin haemorrhages produce the typical dark spots, called 'God's tokens' because they almost invariably herald the victim's death.

This gruesome picture describes classic bubonic plague lasting four to five days, but during epidemics many people died just hours after its onset. This can happen if a flea injects bacteria straight into a small blood vessel, when the disease progresses so rapidly that the victim is dead before the buboes develop. In other cases *Y. pestis* escapes from the buboes and settles in the lungs, causing pneumonia. Then sufferers cough up bloody sputum heaving with bacteria, so switching the transmission route to airborne droplets. Anyone inhaling these droplets develops primary pneumonic plague, which is rapidly and universally fatal (Figure 4.3). So, unlike classic bubonic plague, pneumonic plague spreads directly from person to person. In most epidemics up to a quarter of cases are pneumonic, but there have been entire epidemics of pneumonic plague, the most recent in Manchuria in 1910 and 1920. Yet although the

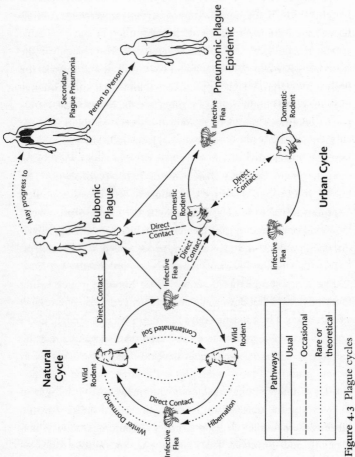

Figure 4.3 Plague cycles

Source: Neal R. Chamberlain, Ph.D., A.T. Still University/KCOM.

mortality is devastatingly high, the droplets generated are heavy so the microbe can only spread a few metres from the sufferer, and as the disease is so rapidly fatal these epidemics generally remain localized and are relatively easy to control.

Most scholars agree that the Black Death as well as the following plague cycles that lasted for 300 years, ending with the Great Plague of London, were epidemics of classic bubonic plague caused by *Y. pestis*. But since the bacterium was not identified until 600 years after the Black Death, this is mainly speculation based on eyewitness accounts, particularly of the characteristic buboes. Recently a few scientists and historians have questioned this assumption, making the points that buboes are not unique to the plague, and that at the time of the Black Death people were in no doubt that sufferers were highly infectious.[11] There are many first-hand reports attesting to this; the following is from Guy de Chauliac, physician to Pope Clement VI, written as the Black Death hit Avignon in 1348:

It was so contagious, especially that accompanied by spitting of blood, that not only by staying together, but even by looking at one another, people caught it, with the result that men died without attendants and were buried without priests. The father did not visit his son, nor the son the father. Charity was dead and hope crushed.[12]

This description could of course refer to the pneumonic form of the disease as the 'spitting of blood' tends to suggest, but the quarantine and isolation measures put in place to prevent direct person to person spread would have had no restraining effect on rats and their fleas, yet on occasions they did seem to restrict the epidemic. Add to this the fact that even the most detailed eyewitness accounts fail to mention a single dead rat, and we certainly have a basis for closer investigation.

The now famous events in the village of Eyam in Derbyshire, England, which was struck by plague in September 1665, serve to illustrate the widely held belief in isolation measures. According to popular legend *Y. pestis* arrived in the village in a box of cloth (presumably containing an infected flea) delivered from London (where plague was raging). Itinerant tailor George Viccars, who opened the box, rapidly fell ill and died of the plague. Starting with this single case an epidemic spread throughout the village, and although it virtually disappeared in the winter, it reappeared in the spring of 1666 with increased ferocity. By this stage all those with the means to do so had deserted the village, and Reverend William Mompesson, Rector of the village church, persuaded those remaining to isolate the village in the hope that if no one entered or left the epidemic would be contained, saving the surrounding population. So from May to December 1666 the villagers were entirely cut off from the outside world, the only points of contact being boundary stones above the village where their grateful neighbours left food and medical supplies. The people could only wait and watch as the plague decimated families, created orphans and divided lovers. The epidemic reached its peak in August and was over by December; in the course of the year it wiped out 259 of the 350 villagers. No plague was reported that year in nearby villages, so although isolation did not prevent the microbe from its relentless spread through Eyam, it was apparently successful in avoiding spread further afield.

Those who doubt that this epidemic was caused by *Y. pestis* argue that isolation would not have prevented rats from carrying the microbe to other villages, but since Eyam was in a remote spot it was perhaps more likely to have spread to the surrounding villages by infected fleas carried on humans or in goods, just as it is supposed to have arrived in the village. Mompesson, who lost his wife Catherine in the epidemic, kept meticulous records of the destructive course of the microbe, and today's History Trail through the village is a poignant reminder of its sad past. An accurate household by household reconstruction of the whole episode suggests to some that the microbe spread directly between people with an incubation period of around thirty-two days, and that victims were infectious for the last eighteen days. Then followed the illness lasting five to seven days and ending in death or recovery.[13] If this scenario is correct then the disease the villagers suffered was certainly not the plague.

Other arguments that convince some experts that whatever caused the ravages of the Black Death and subsequent outbreaks was not *Y. pestis* relate to the black rat and its resident fleas. Since there were no plague foci in Europe in the Middle Ages (and still are not) the microbe had to be imported at the start of the pandemic and then maintained in a resident rodent population. And since brown rats only colonized Europe in the eighteenth century, black rats were the only rodent capable of carrying the microbe. However, they argue that because black rats require warmth they would have been uncommon anywhere beyond Southern Europe. So although they arrived in Northern European ports as part of a ship's cargo, the cold winters of the Little Ice Age stopped them from spreading far inland. But others argue that the Black Death clearly spread throughout Europe along trade routes, the very paths used by black rats as they followed shipments of grain inland from the coast. Also black rat skeletons have been found in the sites of Roman encampments in Britain, including York in the north,[14] but of course this does not prove that they were widespread.

Perhaps a more convincing argument against the Black Death being caused by *Y. pestis* is that rat fleas require a minimum temperature of 18°C for their breeding cycles. So some say that in medieval Northern Europe fleas would only be active in the summer months and could not have transmitted the microbe in winter. Taking this into account, they argue that the disease could not possibly have engulfed the whole of Europe in just three years if flea-borne *Y. pestis* was responsible. The rat/flea argument is most convincing in relation to Iceland, which is said to have been rat-free at the time. Nevertheless devastating episodes of plague occurred in Iceland in 1402 and 1494, carried there by ship from Britain and spreading over the whole island, not even halted by the cold Arctic winters.

One final argument cites the death rate during the pandemic as being too high for genuine plague. Most experts agree that the Black Death killed 30–40 per cent of the population, and in some outbreaks the death rate was as high as 60–70 per cent. This, they say, is higher than that recorded in more recent epidemics where at most 2 per cent of the population died, and far too high for classical plague. However, in a genuine plague outbreak the death rate will vary according to the number of cases of the invariably fatal pneumonic form (spread from person to person).

These pieces of evidence seem to cast reasonable doubt on the Black Death being caused by *Y. pestis* and clearly this controversy will rage on until some definitive proof comes to light. This is likely to be from molecular analysis of human remains, in theory just the kind of all or nothing test that is needed to clear up the mystery. Already a group of French scientists claim to have found *Y. pestis* DNA in the dental pulp of teeth from skeletons thought to be from the Black Death in Montpellier in the South of France.[15] However, others working on teeth taken from mass graves in Northern Europe, including plague pits at Smithfield and Spitalfields in London, fail to find any evidence of *Y. pestis* infection.[16] They disregard the French finding, suggesting that it merely represents contamination of the samples. Clearly we will have to wait a little longer to find out the truth, but in the meantime, if *Y. pestis* did not cause the Black Death then what was it that caused such devastation?

Experts who deny that the Black Death was caused by *Y. pestis* suggest that the pandemic could have been caused by an ancestor of this microbe that behaved very differently from the one we know today, or possibly by anthrax, but they tend to favour an as yet unidentified haemorrhagic fever virus resembling the Ebola virus.[17] They speculate that this Ebola-like virus could have been an ancient zoonotic infection of early humans acquired from other primate species in Africa. They further suggest that it caused the plague of Athens, the Justinian plague and other epidemics around this time, then found a niche in early Middle Eastern civilizations before bursting out to cause the Black Death. These claims may seem far-fetched, particularly with no concrete evidence to go on, but if true, then why was this virus restricted to Europe for 300 years after the Black Death at a time when true bubonic plague was raging in Asia and North Africa? Since there must have been many other microbes circulating during the 300 years of the plague pandemic, perhaps a mixture of epidemics could explain the discrepancies highlighted by those who are sceptical of *Y. pestis* being the sole culprit.

Marseilles in Southern France saw the final outbreak of plague in Western Europe in 1720, and so the last unanswered question about the pandemic, whatever its microbiological nature, is why it disappeared after the Renaissance plague. There are many suggested reasons for the plague's demise, but none are particularly convincing on their own. Firstly, 300 years is quite long enough for genetic resistance to build up in a population and there is some evidence that a gene now carried by around 1 per cent of Europeans today (that inadvertently protects them from HIV infection) may have conferred resistance against the plague.[18] However, *Y. pestis* itself shows no signs of reduced virulence; the 1720 epidemic in Marseilles killed between a third and a half of the population, and when an outbreak struck San Francisco in 1907, seventy-eight of the 160 victims died. This high virulence is probably because *Y. pestis* primarily lives in rodents and is unlikely to adapt to humans as an occasional, dead-end host. Indeed the Indian Plague Research Commission found that only microbes that can grow rapidly and reach very high numbers in a flea's stomach cause the blockage necessary for transfer to humans, so this in itself would select for highly virulent strains.[19]

Temperature change coincident with the beginning of the Little Ice Age around 1450 probably also had a deleterious effect since the plague, whatever its nature, generally struggled to survive the winter in Northern Europe. If this was because rats and their fleas need warmth then the replacement of houses of wood and thatch with brick and tiles, which occurred around this time and was hastened in London by the Great Fire of 1666, must have deprived them of their warm nesting sites. And, indeed, by the eighteenth century the black rat had been mainly supplanted by the hardier brown rat. These changes, combined with the improved general health and nutrition of the population, are the best explanation that can be offered for the end of the ferocious killer in Europe.

Worldwide the Black Death killed an estimated 25 million, and in England alone the death toll was some 1.4 million—around a third of the population. Obviously such a loss in just three years had a profound effect on the survivors and several reminders of the devastating event can still be found in our literature. The famous children's nursery rhyme 'Ring-a-ring of roses' and Robert Browning's poem 'The Pied Piper of Hamelin' are both said to have been inspired by the plague, while Giovanni Boccaccio's collection of 100 novellas in the *Decameron*, beginning with a graphic description of the plague, was written in 1348 as the epidemic raged in Florence.

Whether or not the pandemic actually changed the course of history is still much debated. Of course wherever the pandemic hit chaos ensued, with the fear of infection itself becoming infectious. Even epidemics in modern times, such as HIV, unleash a sequence of terror and panic, denial, anger and a search for scapegoats, the illogical apportioning of blame, low morale and apathy, before the eventual acceptance of the inevitable. And this is in the light of our knowledge of microbes and how to combat them; how much more frightening then an epidemic that strikes out of the blue for no apparent reason. The rich and healthy fled, the dead were left to rot, thieves and cheats took their opportunities, fields were unploughed and cattle untended. But on the whole when the worst was over survivors did what they could to carry on.

Most historians agree that the Black Death, coming at a time of deprivation and famine, accentuated and accelerated the social and economic changes that heralded the modern era.[20] Much has been written about the break-up of the feudal system, but whereas some maintain that the Black Death was the direct trigger for it others believe that the changes were well in progress before the ravages of the Black Death. Whatever the truth of it, the surviving peasants certainly benefited. With the dramatic fall in population that did not recover for 300 years, they suddenly found themselves with more land, and their services in high demand. At the same time overproduction of food led to a fall in prices and a rise in living standards. But with the plague finally gone other microbes rose to prominence, seeming to fill the niche that it had so recently vacated. In particular, the crowd diseases continued to take their toll, and most feared among them was smallpox.

Smallpox

The deadly smallpox virus (*Variola major*), which kills around a third of those it infects, was one of the earliest human zoonotic infections, but exactly how, when and where it transferred its allegiance to man is a controversial issue that will probably never be resolved. Some suggest that smallpox is the human form of monkeypox, acquired by our ancient ancestors in equatorial Africa, and others think that it evolved from cowpox at the time of domestication of cattle. But recent molecular analysis shows that *V. major's* closest relatives are the viruses that cause camelpox and gerbilpox, suggesting that all three arose quite recently from a common ancestor.[21] Like the measles virus, the smallpox virus could only become a purely human pathogen once populations were large and dense enough to sustain it, and among the earliest civilizations likely to provide this niche are those based on irrigation farming in the Euphrates, Tigris, Nile, Ganges and Indus river valleys. Camels and gerbils were both common in these regions, so perhaps an ancient pox virus of rodents jumped to humans and camels some 5,000–10,000 years ago. There is certainly some evidence of smallpox in these centres from early times. Sanskrit medical texts from India mention a disease which sounds like smallpox, indicating that it was probably known there from around 1500 BC, and descriptions in ancient Egyptian papyri written between 3730 and 1555 BC are also suggestive of smallpox.[22] But the most persuasive evidence comes from three Egyptian mummies dating from 1570–1085 BC, which bear skin lesions reminiscent of smallpox. And in samples taken from the lesions of one of these—the mummy of King Ramses V, who died suddenly in 1157 BC while still only in his early thirties—pox-like viruses have been spotted under the electron microscope.[23]

At this time Europe was too sparsely populated to maintain smallpox permanently, so although epidemics may occasionally have spilled over from North Africa, it was not until the rise of the Greek Empire that the microbe could extend its territory in this direction. Indeed the plague of Athens in 430 BC (Chapter 3) could well have been its debut in Europe, and from then on it circulated continuously, particularly in the most populous parts of France and Italy. However, in Britain, where the population remained thin and scattered, smallpox was probably just an occasional visitor, crossing the English Channel from continental Europe from time to time until, arriving with the invading Normans in 1066 and the returning Crusaders of the twelfth and thirteenth centuries, the virus finally managed to establish a permanent base.

Bishop Gregory of Tours gave one of the earliest descriptions of smallpox during an epidemic that hit France and Italy in AD 580:

Such was the nature of the infirmity that a person, after being seized with a violent fever, was covered all over with vesicles and small pustules . . . The vesicles were white, hard and unyielding, and very painful. If the patient survived to their maturation, they broke, and began to discharge, when the pain was greatly increased by the adhesion to the clothes of the body . . . Among others, the Lady of Count Eborin, while laboring under this pest, was so covered with the vesicles, that neither her hands, nor feet, nor any part of the body, remained exempt, for her eyes were wholly closed up by them.[24]

The smallpox virus spreads through the air, is inhaled by its victims, and once inside it multiplies in lymph glands. Then it accesses the bloodstream and targets all the major organs. Here it spends two weeks multiplying until, as thousands of viruses swarm into the blood, the first signs of the illness appear. Starting with sore throat, headache, fever, and general aches and pains which might be caused by any number of microbes, it is not until four days later that the characteristic rash appears and the feared diagnosis is confirmed. The disease varies from deadly haemorrhagic smallpox where bleeding causes the pocks to turn black and coalesce, to the more common form with the pocks progressing through a series of changes as the body's immune system assaults these virus factories. Beginning as discrete red spots, the pocks first fill with fluid then turn from silver to yellow as pus accumulates. On the eighth day the pocks start to rupture, discharging their virus-laden contents, and then drying into scabs. These are eventually shed, all too often leaving unsightly scars, and sometimes causing blindness. During the long course of the illness sufferers generally remain alert but in excruciating pain, not only from the pocks but also from damaged internal organs and suppurating mouth and throat ulcers. The virus is shed in mucus droplets from these ulcers, but being relatively heavy the droplets do not travel far and mainly threaten household contacts. Given the right conditions the virus can survive for a long time in the environment, and since both pock fluid and scabs contain thousands of viruses, these can lurk in dust, or be carried away in clothing or blankets.

R_0 for smallpox is 5–10 and on average the virus spreads to around half of the non-immune contacts. This may sound bad enough, but compared with measles (with an R_0 value of 15 and spread to 90 per cent of contacts) smallpox is the less successful spreader. Although both viruses probably first jumped to humans in the early farming era, the smallpox virus is much less well adapted to humans, and still highly virulent. This striking difference is explained by their genetic make-up; whereas measles has an RNA genome with an inherently high mutation rate, the small-pox virus is a stable DNA virus that will take much longer to adapt to humans. Interestingly, a mutant virus, *V. minor,* which appeared in South America in the early 1900s, caused a milder form of smallpox with a fatality rate of around 1 per cent. This may have been a human-adapted strain which, given time, might have replaced more virulent strains, but since the virus has now been eliminated from the wild we will never know.

As the smallpox virus is not a highly efficient spreader, and in addition there is no silent shedding during the incubation period and no human or animal can harbour a secret store of virus, you begin to wonder how it ever created such havoc around the world, killing at its peak around 400,000 Europeans every year, and an estimated 300 million worldwide in the twentieth century alone. But the virus could not have obtained such a stranglehold if it had not been aided and abetted by humans, and again it was our lifestyle that was instrumental in assisting its spread.

Smallpox is a classic crowd disease that reached its peak as urbanization, followed by industrialization in the late eighteenth century, accentuated the differential between urban rich and poor. Charles Dickens, writing in the nineteenth century, elo-quently documented the plight of the poor in London, where whole families often shared a single room in cold, damp, infested tenements with no ventilation or means of waste disposal. And with no organized support for the needy, the filthy streets thronged with the homeless and destitute. As with almost any infectious disease at the time, the only known preventive meas-ures for smallpox—isolation of cases and fleeing the area—were open solely to the rich, so the poor were left to bear the brunt of an epidemic. In spite of this, the smallpox virus was so ubiquitous that the rich and famous could not escape entirely, and a glance at the family trees of European royalty is enough to convince you that smallpox changed the course of history on many occasions by picking off reigning monarchs and their heirs. In England, Elizabeth 1 had been queen for only four years and was enjoying herself at Hampton Court palace when the microbe struck. She sent her doctor packing when he mentioned the possibility of smallpox, but he had to be hastily recalled when later the same day she developed the characteristic rash and sank into a coma. With no heir, Europe in turmoil and Mary Queen of Scots recently returned from France and now on her very doorstep, this was not a convenient time for the queen to die. Fortunately, she recovered with her beauty unscathed to reign over England for another forty-one years. But others were not so lucky.

As the centuries passed the grip of the virus got steadily tighter. With the plague virtually gone from Europe by the end of the seventeenth century, and with leprosy on the decline, smallpox became the most common killer, accounting for over 5 per cent of all deaths in London. Again the royal family did not escape; indeed when Charles II was recalled from France in 1660, eleven years after his father's execution, everything looked well for the Stuarts. But within a generation they were gone—wiped out by the deadly virus. Charles II lost his brother Henry and sister Mary to the virus, and with no legitimate offspring of his own he was succeeded by his brother James II. James had no living sons, one having died of smallpox, so when his unpopular Catholic faith caused him to relinquish the throne three years later it passed to his daughter Mary II and her husband and cousin William of Orange. Shortly after their succession the still childless Mary died of smallpox, so William, who survived smallpox as a child but lost both his parents to the disease, ruled alone for eight years.

He was succeeded by Mary's sister Anne, whose only son died of smallpox, so her death heralded the end of the Stuart line. And the British monarchy was not the only one in Europe to suffer from the smallpox virus. Within eighty years of the demise of the Stuarts, Emperor Joseph I of Austria, King Luis I of Spain, King Louis XV of France, Queen Ulrika Eleonora of Sweden and Tsar Peter II of Russia all died of the disease.

Smallpox epidemics increased in frequency and ferocity right through to the nineteenth century when the pattern was finally broken, first by the advent of inoculation and then vaccination (see Chapter 8). But before that the virus, along with hundreds of other microbes, went global, beginning its cycle of killing in the totally naïve population it encountered in the New World.

5

MICROBES GO GLOBAL

A s the sea level rose at the end of the last Ice Age, the Bering Strait land bridge linking Siberia and Alaska submerged and America became an island. The descendants of the Mongoloid people who crossed the bridge some 14,000 years ago were isolated, cut off from the inhabitants of the Old World and their microbes. In previous chapters we have followed the evolution of infectious disease microbes in the Old World; now we cross the Atlantic to catch up with the Native American people in the New World.

By the time contact between the Old and New Worlds was re-established in the late fifteenth century there were around 100 million Native Americans, some living in scattered communities, but with two great flourishing civilizations, the Incas in Peru and the Aztecs in central Mexico, each containing some 25–30 million people. Despite the fact that both these populations far exceeded the size and density needed to sustain the acute infectious diseases familiar to Eurasians, none of these existed in the Americas before the European invasion. So descendants of the hunter-gatherers who crossed from Siberia were familiar with the ancient persistent microbes we all inherited from our ape-like ancestors, such as the herpesviruses, parasitic worms (Chapter 3) and possibly the skin disease yaws (see pages 128–9), but had never met the microbes responsible for the 'new' acute infectious diseases.

The reason for the absence of crowd disease microbes in the Americas is probably quite straightforward; in the Old World these microbes jumped from domestic animals (see Chapter 3), but, in the Americas, where the wild game was rapidly depleted by hunter-gatherer bands, very few suitable species remained to be domesticated. In fact throughout both North and South America Native Americans only farmed turkeys, ducks, guinea pigs, llamas and alpacas, none of which are traditional herd animals likely to donate microbes to humans. Even dogs, which were probably domesticated from wolves some 60,000 years ago, and most likely accompanied hunter-gatherers across the land bridge, apparently did not carry microbes capable of making the transfer.

The meeting of the New and Old Worlds in the fifteenth century, driven by the desire to find and exploit new territories, had potential benefits on both sides. For Eurasians calorie-packed potatoes and maize, and vitamin-rich chilli peppers and tomatoes helped stave off deficiencies and starvation, while Old World domesticated animals supplemented the mainly vegetarian Native American diet and overcame its vulnerability to famine. But with trading of goods and people there also came the inevitable exchange of microbes, and this was not quite such a fair two-way process. The direction of spread was overwhelmingly from the European invaders to the indigenous population and the result was disastrous for Native Americans. Once again we witness how rampant microbes can be when let loose among people who lack genetic resistance honed by centuries of exposure. Those that by now caused fairly mild disease in Eurasians produced devastating epidemics among naïve American populations, even, on occasions, wiping out whole communities.

So for Native Americans, the arrival of Christopher Columbus in 1492 represented a watershed between a life relatively free from acute infectious microbes and one of being slain by literally dozens of them. What happened was an action replay of events in Eurasia thousands of years earlier, but now with the fast-forward button in operation. Back then microbes were obliged to wait for each community to reach a critical size and density and for trading routes to open up before they had opportunity to flourish, but from the fifteenth century onwards microbes that survived the Atlantic crossing poured into the New World and took the native people by storm. To the experienced Eurasian immune system these microbes ranged in severity from smallpox and diphtheria, still lethal in around 10 per cent of cases, through measles, scarlet fever and whooping cough, to the milder flu, mumps, German measles and the common cold. But to the naïve immune system almost all spelled disaster; the result was a devastating 90 per cent drop in the Native American population over the next 120 years.[1] Whole tribes were wiped out, their culture and languages lost for ever. Indeed, in Hispaniola (now Haiti), home to around 8 million Native Americans when Columbus arrived, not one remained some forty years later. In the words of a Maya Indian of Yucatan who wrote longingly (and perhaps a little idealistically) of the time before the arrival of Europeans:

There was then no sickness; they had no aching bones; they had then no high fever; they had then no smallpox; they had then no burning chest; they had then no abdominal pain; they had then no consumption; they had then no headache. At that time the course of humanity was orderly. The foreigners made it otherwise when they arrived here.[2]

When he first set foot in the Caribbean Columbus found a land ideal for growing sugar cane, with plenty of cheap labour to man the plantations, but by the early sixteenth century European demand for sugar far outstripped the capacity of the Spanish plantations to produce it. Both land and labourers were in short supply, so when the Spanish Governor of Cuba, Diego Velázquez, heard rumours of a rich and thriving civilization in Mexico, he sent Hernando Cortés to investigate. What followed is a familiar story, but worth retelling here as a powerful illustration of how microbes can influence the course of history. Cortés set sail for mainland America with just sixteen horsemen and 600 foot soldiers, and in 1519 landed on the coast of Mexico where he established a base and began gathering information and recruiting allies among the local Indians. This small group then made its way to the Aztec capital of Tenochtitlán where Emperor Montezuma, believing that Cortés was the white-skinned god, Quetzalcóatl, returned to fulfil an ancient Aztec prophecy, welcomed them enthusiastically. But, perhaps because Cortés did not behave in a very godlike manner, relations soon deteriorated and he hurriedly retreated back to the coast, losing over half his men in the process. While he spent most of 1520 recruiting more local Native Americans and preparing for a counter-attack, smallpox struck the Aztec capital. The people, all naïve to the virus, died in their thousands. A Native American later wrote this lament of the smallpox attack:

. . . it spread over the people as great destruction. Some it quite covered on all parts—their faces, their heads, their breasts. There was a great havoc. Very many died of it. They could not walk; they only lay in their resting places and beds. They could not move; they could not stir; they could not change position, nor lie on one side; nor face down, nor on their backs. And if they stirred, much did they cry out. Great was its destruction. Covered, mantled with pustules, very many people died of them.[3]

When Cortés returned to besiege Tenochtitlán in 1521 he added starvation to the devastation wreaked by the smallpox, and the city fell in just seventy-five days.

An epidemic that was probably also smallpox similarly favoured Francisco Pizarro and his troops when they invaded the Inca Empire in 1532. The first smallpox epidemic struck the empire some time in the 1520s, killing around a third of its people and devastating the royal household. Emperor Huayna Capac, the absolute monarch, worshipped by his people as the Sun God, died along with many of his military leaders, governors and family members. But it was the concurrent death of his son and heir, Ninan Cuyuchi, that left the empire in confusion and precipitated a war that ended in the lands being split between two competing factions. Pizzaro with just sixty-two mounted and 106 foot soldiers marched into this disarray in 1532, and despite the 80,000-strong force headed by the Emperor Atahualpa that turned out to greet him when he entered Cajamarca, he had an easy victory. Much to the surprise of Pizzaro and his men, who were unnerved by the mass of troops before them, they captured Atahualpa without the loss of a single soldier. One of Pizzaro's men wrote to the King of Spain:

All of us were full of fear, because we were so few in number and we had penetrated so far into a land where we could not hope to receive reinforcements.

First came a squadron of Indians dressed in clothes of different colors, like a chessboard. They advanced, removing straws from the ground and sweeping the road. Next came three squadrons in different dresses singing and dancing. Then came a number of men with armor, large metal plates, and crowns of gold and silver. So great was the amount of furniture of gold and silver which they bore, that it was a marvel to observe how the sun glinted upon it. Among them came the figure of Atahuallpa in a very fine litter with the ends of its timbers covered in silver. Eighty lords carried him on their shoulders, all wearing very rich blue livery. Atahuallpa himself was very richly dressed, with his crown on his head and a collar of large emeralds around his neck.[4]

Once the emperor was captured the battle was over. Pizzaro held him to ransom, extracting enormous wealth from his subjects in return for his freedom. But when Pizzaro had all the riches he wanted he had the emperor executed.

There are many reasons why these and other Spanish victories over Native Americans were inevitable, not least because most had never seen white men before (those 'monstrous marine animals, bearded men who move upon the sea in large houses'[5]), and, in the case of the Aztecs, they interpreted Cortés' arrival as the fulfilment of an ancient prophesy. Also the effect of an entirely new type of warfare, with charging horses, blazing guns and razor sharp swords, must have been absolutely terrifying to those armed with stone axes, slings, bows and arrows. But still there is no doubt that the devastating microbes that struck Native Americans played their part in the lack of resistance to these early attacks, and the consequent ease and swiftness of the victories of the grossly outnumbered Spanish. The Native American's fate was sealed by smallpox, the demoralizing, disorientating, numbing effect of a sudden outbreak of this disfiguring disease that appeared out of the blue, killing thousands for no apparent reason. Not only did the epidemics, so conveniently timed to aid the Spanish, severely depopulate the Indian fighting forces, deprive them of their leaders and demoralize the ranks, but both the Spanish and the Native Americans believed that the devastating diseases—which killed around a third of Native Americans while sparing the Spanish (most of whom were immune after surviving childhood infections)—were sent by an angry god as punishment for previous misdemeanors. The terrible sequence of events seemed to confirm Spanish superiority and the Americans could only acquiesce.

Epidemics of smallpox were inevitably followed by others as dozens of microbes from the Old World, including measles, German measles, flu, diphtheria, scarlet fever, typhoid, whooping cough, dysentery, mumps and meningitis, crossed the Atlantic, so that within fifty years of Cortés' arrival in central Mexico only one in ten Native Americans survived and the population plummeted from 30 to 3 million.

Deaths from imported diseases among these previously isolated peoples peaked in the sixteenth and seventeenth centuries, and by 1700, dispersal of Eurasian microbes in the Americas, and perhaps also a few American in Eurasia, was complete. Thereafter the pattern of infectious diseases in the New World stabilized into cyclical childhood epidemics that had been established centuries before in Eurasia.

Events similar to these, although not on quite as large a scale, were enacted in many other isolated communities throughout the world, including the Aborigine and Maori people of Australasia, the Pacific Islanders and the Southern African Khoisan people. Their populations were decimated, often mainly by measles, and some never recovered. Once again the people, their cultures and languages were lost for ever.

The Slave Trade

With the massive fall in the Native American population in so short a time, owners of the lucrative European sugar plantations in the Caribbean were crying out for cheap labour, and the solution they found was to import slaves from Africa. This trade in human bodies began early in the sixteenth century, reached its peak between 1640 and 1680, and was outlawed in most countries by 1820. During this period an estimated 12–20 million Africans were transported to the Americas, mostly from West Africa to the Caribbean sugar plantations.

West Africa in the sixteenth century, known as the 'white man's grave', was teaming with deadly microbes, particularly those causing malaria and yellow fever, and it was not long before they arrived in America. These 'new' microbes attacked both Native Americans and Europeans viciously, but malaria tended to spare African slaves because of their genetic resistance to it. So although they were susceptible to other 'European' microbes, African slaves did not die off as rapidly as Native Americans and soon came to outnumber them in many areas, particularly the Caribbean.

Importation of the malaria parasite and yellow fever virus into the Americas was not quite as straightforward as the acute infectious disease microbes had been because both these microbes require mosquito vectors. Malaria parasites can survive for some time in the blood of apparently healthy carriers, so they probably made the crossing from Africa to the Americas many times inside African slaves. However, the microbe could not take hold in the New World until it found a suitable mosquito to transmit it from person to person. It is possible that this too was imported from Africa, but given the speed at which the disease spread, it is more likely that the microbe found local American mosquito species to do the job. By 1650 or thereabouts malaria was endemic in the Caribbean and low-lying tropical mainland areas, and from here it spread throughout the Americas. The parasite was only eliminated from the US in the early twentieth century, and remains a threat in several parts of South America today.

By the mid-1600s yellow fever virus also had a firm foothold in the New World. This virus may just cause a relatively mild flu-like illness (typically with headache, fever, muscle aches, nausea and vomiting), but 5–20 per cent of sufferers go on to develop a fatal haemorrhagic fever. The various names given to the deadly disease graphically portray its main symptoms: 'yellow fever' refers to the jaundice which accompanies liver failure, and the Spanish name, 'vomito negro', denotes the black vomit caused by internal bleeding. But 'yellow jack', coined by British sailors, refers to the yellow quarantine flag flown by yellow fever-stricken ships arriving in port, and not to jaundice.

Yellow fever virus naturally infects monkeys in the tropical rainforests of West Africa, where it is spread among them by mosquitoes that live in the tree canopy and breed in rain-filled tree holes. This natural forest cycle causes no problem to the monkeys, but they serve as a reservoir of the virus, so anyone entering the forest is at risk of being bitten by a virus-carrying mosquito and developing the deadly disease. Forest clearance is a particularly dangerous occupation since tree-felling brings canopy mosquitoes and their breeding pools down to ground level. And once humans are infected the virus can be spread directly among them by mosquitoes in a so-called urban cycle. *Aedes egypti* mosquitoes are particularly well adapted for this task as they like to live alongside humans, feeding off their blood, sheltering in houses and breeding in their water tanks. And once these mosquitoes pick up the virus they carry it for life (one to two months), even passing it on to their offspring.

Unlike malaria, yellow fever sufferers either die or recover completely; there are no healthy human virus carriers. So the virus must have made the trip from West Africa to the Americas on board slave ships with the help of its mosquito vector. These insects could survive the six to eight week transatlantic crossing quite happily by breeding in the ships' water barrels, spreading the virus to passengers and crew along the way. Then on arrival virus-laden mosquitoes flew ashore from quarantined ships, starting an epidemic and frustrating all attempts to control the disease.

Barbados was host to the first reported yellow fever epidemic in the New World in 1647, and as the slave trade increased, the virus spread through the Caribbean islands and South America and then made its way north to ports along the Atlantic coast of the US. At first each epidemic was sparked by a fresh batch of mosquitoes arriving with a cargo of slaves, but soon these insects took up residence in the hot humid areas of the South, where they infected the resident monkey population, and the disease became endemic. The first epidemic in the US was in Philadelphia, the seat of government, where it killed 4,000 and was a major factor in the decision to build the new capital in Maryland.[6] At its peak the virus invaded all the US Atlantic ports from Charleston to Boston, and made its way along the Mississippi River to New Orleans and Memphis, where a massive epidemic in 1878 killed around a tenth of the population.

Along with malaria, yellow fever played a major part in depopulating tropical areas of the Americas, particularly the Caribbean. Native Americans were highly susceptible, and although it was commonly believed that African slaves were resistant to yellow fever as well as malaria, many died in the epidemics. The microbe also took its toll of European settlers and continually frustrated French and British interests in the South. Indeed, Napoleon dreamt of making Santa Domingo the capital of his proposed New World Empire until an outbreak of yellow fever in 1801 decimated the troops he sent to the island to quell rebellious slaves. With thousands dead his plan to move on and occupy New Orleans was in ruins, and he eventually abandoned his American dream entirely and sold the French territory of Louisiana to the

US for 15 million dollars. The yellow fever virus also aborted French attempts to build a canal across the isthmus of Panama. Beginning the project in 1880, they struggled with it for twenty years, but the canal was eventually completed by the Americans in 1913 after the yellow fever had been controlled. The whole enterprise cost over 300 million dollars and sacrificed 28,000 lives.

For a long time no one could understand how yellow fever spread. Opinions were divided between the 'contagionists' who believed that the disease was infectious, pointing out that epidemics always began with the arrival of ships carrying infected passengers. They urged stricter quarantine laws, but the 'environmentalists' argued that quarantine was ineffective and preferred to blame the filthy, unhygienic condition of the ports. In reality both sides were partly right, although neither had an inkling that the main culprit was a virus-carrying insect that could bypass quarantine laws but required hot and humid conditions to breed; the whole infectious cycle took a long time to unravel. As early as 1847 the idea of a mosquito vector for yellow fever was mooted, but after Dr Carlos Finlay, working in Havana in 1881, failed to transmit the disease from patients to volunteers using mosquitoes, the idea was dropped. In fact his experiments probably failed because of timing; the virus only circulates in victim's blood briefly just before the jaundice appears, and it has to incubate inside a mosquito for a week before it can be passed on, so the mosquito feeding times are critical for successful transmission.

The disease raged on until it was forcefully brought to the attention of the US army during the Spanish–American war of 1898 when US troops fighting in the Caribbean lost just 968 men in combat, but over 5,000 died of yellow fever.[7] The army sent a team of four doctors headed by bacteriologist Walter Reed to Havana to investigate. The team agreed to use themselves as the first guinea pigs to test out the mosquito theory, and Reed's assistant, James Carroll, and bacteriologist Jesse Lazear were the first to volunteer. After being bitten by mosquitoes that had fed on yellow fever patients Carroll developed the infection and recovered, whereas Lazear remained well. But then Lazear, the only team member who really believed that mosquitoes were the culprits, was accidentally bitten while collecting blood from a yellow fever patient. This time he came down with severe disease and died twelve days later.

Not deterred, Reed continued the experiments on human guinea pigs until he had shown conclusively that mosquitoes transmit yellow fever. And although the nature of the microbe was still not known, William Gorgas, Havana's Chief Sanitary Officer, set about ridding the city of mosquitoes, and in just three years had routed the virus. With this success, public health doctors firmly believed that yellow fever could be eradicated by mosquito control alone, but after discovering the monkey reservoir in Africa, and finding that several different types of mosquito could spread the virus in the jungle setting, they finally realized that this zoonotic microbe was here to stay. Then prevention became the goal and after the yellow fever virus was isolated in 1927 a vaccine was soon being trialled. This helped to clear the infection from the US, but the disease is still a major health problem in West Africa, where most of the annual 200,000 cases and 30,000 deaths occur.

Following Columbus's epic journey, the main flow of microbes was unequivocally from East to West, but a few may have travelled in the opposite direction. Three new diseases (syphilis, typhus and the English sweats) made their debut in Europe around the time of Columbus's return, but whether the microbes were carried back from the New World is far from clear. The most likely candidate for importation is syphilis, but it is possible that typhus and the English Sweats also arrived from the Americas. About the mysterious English sweats there is little to say; unusually it preferentially struck males of the affluent classes in rural areas in the summer months. For these reasons some suggest that the culprit microbe was zoonotic, perhaps jumping from rats or mice to start an epidemic which was then fuelled by human to human spread. The disease had a high mortality, and raged in Britain and Europe for around seventy years before disappearing entirely.

In contrast to the English sweats, typhus came to stay, but no one knows for certain whether this microbe crossed the Atlantic from East to West or vice versa. Some say it first appeared in Spain around the time of Columbus's return and then spread to Italy, causing an outbreak among French troops battling against the Italians. This suggests an American origin, and is backed up by the fact that American flying squirrels act as a reservoir of the microbe. However, others believe the microbe has infected Eurasians since ancient times and was transported to America along with all the other microbes. But as typhus was only clearly distinguished from typhoid in 1837 it is not possible to prove or disprove these theories, and we will revisit the disease in the next chapter when discussing infections in the nineteenth century, by which time its identity was clear.

Syphilis

Syphilis made its first dramatic appearance on the European stage in 1494. Charles VIII of France had invaded Italy and captured Naples with the intent of claiming the throne when an epidemic struck his troops. Indeed Charles himself is said to have been among the early sufferers; if true, he was the first of a number of monarchs to catch the disease. In the words of the historian of the

House of Burgundy he had 'a violent, hideous and abominable sickness by which he was harrowed; and several of his number, who returned to France, were most painfully afflicted by it; and since no one had heard of this awful pestilence before their return, it was called the Neapolitan sickness.'[8] But Charles certainly suff-ered no long-term consequences, as he died of apoplexy three years later after hitting his head on a door frame.

This new disease was probably instrumental in ending Charles's occupation of Naples and as his army retreated in disarray, soldiers returned to their homelands dispersing the microbe far and wide. The epidemic spread like wildfire throughout Europe, so that by the end of the century it raged from London to Moscow and then proceeded to invade Africa and China. The English called the new disease 'the great pox' to distinguish it from smallpox, but in other countries the names used generally pointed the finger at those who had supposedly caused the disaster. Thus the Italians called it 'the French disease', the French 'the disease of Naples', the Poles 'the German disease', and the Russians 'the Polish disease'. In the Middle East it was named 'the European pustules', in India 'the Franks', in China 'the ulcer of Canton', and in Japan 'Tang sore'.[9] The name 'syphilis' was coined some thirty years later by the Italian physician and scholar, Girolamo Fracastorio, who was himself a sufferer. He wrote a poem (Syphilus sive morbus gallicus) about a young shepherd boy named Syphilus, which refers to the disease:

> He first wore buboes dreadful to the sight
> First felt strange pains and sleepless passed the night
> From him the malady received its name.[10]

In the late 1490s physicians were unanimous in thinking that they were dealing with a new disease, for although they were familiar with sexually transmitted genital ulcers they had never seen a generalized disease spread in this way. Compared to the syphilis we know today the disease they described was much more severe and rapidly progressive, with skin, mouth, throat and genital ulcers accompanied by high fever, intense head, bone and joint pains and a variety of rashes, which as the name great pox infers, could be so florid as to mimic smallpox (Figure 5.1). Sufferers were extremely sick and often died at this early stage, an outcome almost unknown today.

From the following rather quaint English description written in 1547, it is obvious that the sexual mode of transmission was widely appreciated from early on:

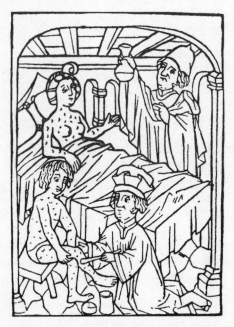

Figure 5.1 Title page from Bartholomew Steber's *Syphilis*, 1497 or 1498

The cause of these impediments or infyrmytes doth come many wayes, it maye come by lyenge in the sheets or bedde there where a pocky person hath the night before lyenin, it maye come with lyenge with a pocky person, it maye come by syttenge on a draught or sege where as a pocky person did lately syt, it may come by drynkynge oft with a pocky person, but specially it is taken when one pocky person doth synne in lechery the one with another.[11]

Syphilis is caused by the highly motile, corkscrew-shaped spiro-chaete bacterium, *Treponema pallidum,* that is transmitted either sexually or from mother to unborn child through the placenta. In adults the first sign of the disease is a relatively painless genital ulcer, or chancre, from which the microbe accesses the bloodstream and invades internal organs. This produces the symptoms of secondary syphilis, including fever, enlarged glands, skin rashes and mouth and genital ulcers, all of which resolve in a few weeks. Then begins a latent period of up to twenty-five years during which the microbe hides away but continues to grow, producing internal ulcers called gumma, that slowly and inexorably destroy the surrounding tissues. At the same time the microbe may invade blood vessels and the brain, causing the array of problems of tertiary syphilis that include heart attacks, strokes, blindness, deafness, personality changes and loss of intellect. *T. pallidum* is penicillin sensitive so these days the disease can be cured in its early stages, but the tissue destruction of the later stages is, of course, irreversible.

No one disputes the fact that the arrival of syphilis in Europe coincided with the return of Columbus's men from Hispanola in 1493, but whether the two events are connected is hotly debated. The traditional story tells of *T. pallidum* disembarking with the sailors in Lisbon, and there was no doubt in the mind of Ruy Diaz de Isla, the Spanish physician who treated Alonso Pinzon, captain of the Pinta, and some of his crew on their return, that this was the case. He later wrote:

And since admiral Don Christopher Columbus, who had relations and congress with the inhabitants of this island during his stay, discovered this island and since the disease is naturally contagious, it spread with ease, and soon appeared in the fleet itself. [12]

On their return some of Columbus's crew immediately joined Charles VIII's army of mercenaries in their attack on Naples in 1494, so in the chaos of the campaign they could well have sparked an epidemic among the troops and camp followers.

Until recently, the American origin of *T. pallidum* was backed up by finding the telltale signs of syphilis in the bones and teeth of pre-Colombian American skeletal remains, although this was not found in pre-1492 European skeletons. However, in the mid-1990s signs suggestive of syphilis were reported in skeletons of monks from an English monastery, dating from between 1300 and 1450—well before Columbus sailed. [13] But gross examination of bones cannot provide definitive proof of syphilis since the deformities it produces are indistinguishable from other diseases, particularly leprosy and yaws. DNA analysis can differentiate between the microbes that cause syphilis and leprosy, but at present this test cannot distinguish between yaws and syphilis because the microbes are almost identical.

Yaws is one of a group of chronic skin infections caused by *T. pallidum*-like spirochaetes that flourish in conditions of poor personal hygiene. These microbes cause deep gummatous ulcers on skin and mucous membranes, but in the late stages of the disease the microbe may attack bones and joints. Yaws-like infections, variously called bejel, pinta, bubas and framboesia in different parts of the world, mainly spread among children by close contact. Yaws is presumed to be an ancient human disease, but it is impossible to untangle its history in Europe because in medieval times chronic disfiguring skin diseases were all lumped together under the name of 'leprosy' and the stricken were treated as social outcasts, often confined to leper colonies. But in the aftermath of the Black Death leprosy began to decline and although the exact reason for this is unclear, with up to 40 per cent of the population dead from the plague, living quarters were inevitably less crowded and there was more food, fuel and clothing to go round. So in general living standards improved, and as the Little Ice Age took hold maybe the need for extra warmth was satisfied more by clothes and fires than by huddling together in a shared bed as of old. For microbes like the yaws spirochete that spreads by close skin contact and is too fragile to survive outside the human body, this could have spelled disaster, but some speculate that at this stage it simply switched its mode of transmission from direct skin contact between children to the more intimate sexual contact between adults, so taking on the guise of syphilis.[14]

This theory ties in neatly with the absence of syphilitic bone lesions in pre-Columbian Europe, and assuming that the change in transmission coincided with the French–Italian war of 1494, then it could have been the culprit of the virulent 'new' disease infecting the troops. Now that the entire DNA sequence of *T. pallidum* is known,[15] and some differences between it and the yaws spirochete have been identified, there is hope that this controversy will soon be resolved.

T. pallidum, which caused such severe disease in the first pandemic, soon lost virulence and by the seventeenth century the disease had assumed its present-day characteristics. So although it caused an outcry at the time, syphilis had only a minimal effect on human history. Those interested in its overall influence on world affairs focus on individuals whose bizarre behaviour may be attributed to the effects of tertiary syphilis on the brain. Suggested sufferers include King Henry VIII of England, whose failure to produce a live male offspring with his first wife, Catherine of Aragon, may have been the work of *T. pallidum*. And since his desire for an heir induced him to marry six times, a feat that the Pope would not countenance, *T. pallidum* could be held responsible for the break with the Church of Rome and the establishment of the Church of England. Another possible victim is Ivan the Terrible, first Czar of Russia, whose famous megalomaniac behaviour in later life has been attributed to neurosyphilis. Perhaps, too, the microbe had a more widespread influence in changing moral attitudes in Europe and causing the rise of Puritanism.[16]

Cholera

The British East India Company was set up in 1600 to exploit the market for luxuries such as silks and fine cottons, indigo, sugar and spices. Trade flourished, but after the Indian Mutiny in 1857, the British Government took control and spent the next 90 years trying to annex Indian states to the British Crown.

Prior to British rule in India, the cholera microbe was confined to its natural homelands in the Bay of Bengal, where the great River Ganges flows into the Indian Ocean. There have been seasonal outbreaks of cholera here since time immemorial, often linked to crowds gathering for Hindu pilgrimages and festivals. But when the British opened up a network of trade routes, and troop movements escalated, they gave the cholera microbe its chance to move on to the world stage.

Cholera is caused by *Vibrio cholerae*, a comma-shaped bacterium that lives in water and swims by lashing its whip-like flagellum. It is generally spread between humans by faecal contamination of drinking water, and once swallowed vibrios must run the gauntlet of the acid environment of the stomach, where many are inactivated before reaching their destination in the small intestine, so the infective dose for cholera is high. Those that survive attach to the gut wall and produce a potent toxin composed of A and B subunits. While B tethers the toxin firmly to the gut lining, A is injected into the lining cells, blocking their power to absorb water from the gut and also releasing stored body fluids containing essential sodium and potassium ions. This causes a massive outpouring of fluid into the gut, together with its sudden forceful expulsion from both ends and severe, painful spasms of the abdominal muscles. The resulting diarrhoea can be so profuse that the mucus flecked fluid effluent, aptly named 'rice water stool', can exceed 1 litre every hour. In no time at all sufferers are dehydrated and if nothing is done to replace the fluid irreversible shock sets in, causing death in a matter of hours. Without treatment the death rate for severe cholera is around 50 per cent.

Vibrio cholerae can cause huge epidemics anywhere that poor sanitation gives it access to water supplies, and in the nineteenth century this was the situation in almost every large city in the world. So a cholera pandemic was a disaster waiting to happen when in 1817 unusually heavy monsoon rains in Bengal caused extensive floods and crop failures and precipitated a severe local cholera outbreak. This happened to coincide with British troop movements in the area and the epidemic mushroomed. Troops and traders dispersed the microbe throughout the Indian subcontinent and the West as far as Southern Russia, while ships carried it much further afield. Eventually the pandemic reached China, Japan, South-East Asia, the Middle East and the East African coast, before dying out in early 1824. In the words of the Marquis of Hastings, whose army division was camped in Vindhya Pradesh at the time:

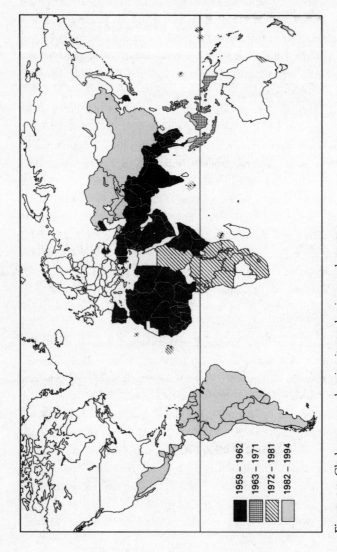

Figure 5.2 Cholera: a map showing its pandemic spread, 1959–94

Source: Adapted from R. L. Guerrant, 'Lessons from diarrheal disease: demography to molecular pharmacology, *J Infec Dis* 169(6): 1206–18, 1994, by permission of University of Chicago Press.

1959 – 1962
1963 – 1971
1972 – 1981
1982 – 1994

The march was terrible for the number of poor creatures falling under the sudden attacks of this dreadful infliction, and from the quantities of bodies of those who died in wagons and were necessarily put out to make room for such as might be saved by the conveyance. It is ascertained that above 500 have died since sunset yesterday.[17]

To date there have been seven cholera pandemics, with *Vibrio cholerae* travelling progressively further from its homeland (Figure 5.2): during the second (1826–1832) it reached Europe and then jumped across to Canada in a boatload of Irish immigrants; the US was hit in the second and more extensively in the third (1852–1859); and South America was badly affected during the fifth pandemic (1881–1896). Between each of these pandemics the microbe kept its foothold in the Ganges Delta. The seventh pandemic began in 1961 and is still ongoing. Although the poor were always the worst hit by cholera, the disease was by no means restricted to them and epidemics could strike anywhere. It is not surprising that the threat of an outbreak caused alarm and panic when a person could be perfectly healthy one day and die a horrible death the next, literally shriveling to a corpse before the eyes of family and friends, who could only look on helplessly.

Recurrent epidemics of cholera in Europe and the US in the nineteenth century prompted governments to set up Health Boards to oversee these outbreaks, but it was not until the shrewd detective work of the British doctor John Snow, during the London epidemic of 1854, that drinking water was identified as the source. Snow was an anaesthetist by trade but he had a long-standing interest in cholera, sparked by witnessing an epidemic in 1832 while working as a medical apprentice in the mining town of Newcastle in north-east England. Most people at the time believed the disease resulted from exposure to miasmic vapours, but Snow was sure that the cholera 'poison' did not pass through the air, commenting that 'there are a number of facts that have been thought to oppose this evidence: numerous persons hold intercourse with the sick without being affected, and a great number take the disease who have no apparent connection with other patients'. After examining the gut of dead victims, he noted: 'local affection of the mucous membrane of the alimentary canal'. He concluded that:

the disease must be caused by something which passes from the mucous membrane of the alimentary canal of one patient to that of another, which it can only do by being swallowed; and as the disease grows in a community by what it feeds upon, attacking a few people in a town at first, and then becoming more prevalent, it is clear that the cholera poison must multiply itself by a kind of growth... this increase taking place in the alimentary canal. The instances in which minute quantities of the ejections and dejections of cholera patients must be swallowed are sufficiently numerous to account for the spread of the disease.[18]

Snow then set about proving this controversial theory that cholera was a water-borne disease. When an epidemic struck London in 1854 he meticulously mapped its spread from case to case. At the time London's sewage was deposited en masse into the River Thames and those citizens with a piped water supply were served by several different water companies. All drew their water from the Thames, but at varying points along its course—some above and others below the sewage effluent. Snow checked the water source used by each cholera victim and found that households using water from the Southwark and Vauxhall Water Company, drawn downstream of the city, suffered nine times more cholera than those supplied by the Lambeth Company, which drew water from further upstream. He followed this by charting cholera cases in his local London parish of Soho and noted that most affected households got their drinking water from the Broad

Street pump. He famously persuaded the parish council to inactivate the pump by removing its handle so that users had to go
elsewhere for their water, and sure enough the number of cholera
cases in the area served by the pump dropped dramatically. It turned
out that the well water was contaminated with sewage from a
nearby house where a child had recently suffered from cholera.

On the basis of his observations Snow suggested simple measures
to prevent cholera, such as hand washing, boiling drinking water
and decontaminating bed linen, but unfortunately he died three
years later from a brain haemorrhage at the age of forty-five and
never saw his theory fully accepted or these measures implemented.
However, his work encouraged the germ theorists, and in 1883
when a cholera epidemic in Calcutta and Bombay spread to Egypt,
Robert Koch headed for Alexandria with a team of microbiologists
to identify the causative microbe. They found dead victims' guts
teaming with comma-shaped bacteria and then, having followed the
epidemic to Calcutta where they isolated the microbe, announced
their findings in 1884.

A vaccine for cholera was ready by the late 1880s, but although
useful for travellers to endemic areas, it was too expensive and its
effects too short-lived to be of any help for those living in these
regions, so the high death tolls continued until simple rehydration
of victims by intravenous infusion revolutionized treatment and
death rates plummeted. Rather surprisingly, *Vibrio cholerae* can
survive independently of humans or any other animal reservoir.
Between epidemics vibrios live quite happily in the estuarine
waters of the Ganges Delta, where they are part of the normal
aquatic flora. They cling on to, and live off, the chitinous surface
of plankton such as diatoms, shellfish, arthropods and their molts.
During the cyclical algal blooms *Vibrio cholerae* undergoes a population explosion along with its host plankton, which increases the

likelihood of it infecting humans, sparking an epidemic. But things are not quite that simple because most estuarine *vibrios* cannot produce the toxin that causes the symptoms of cholera and so they are harmless. The cholera toxin gene is in fact carried by a phage virus that targets the *vibrio*, so only those *vibrios* infected by this virus are converted into a toxigenic strain. To date pandemics have resulted from just two phage-converted toxigenic *Vibrio cholerae* strains, the famous 01 and 0139 (Figure 5.3).

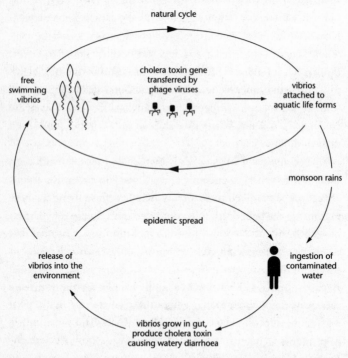

Figure 5.3 Cholera: the natural and epidemic cycles of *Vibrio cholera*

There are many different types of phage viruses in the estuarine waters of the Bay of Bengal, where they live in balance with their host bacteria. As we have seen infection of *Vibrio cholerae* with one of these conveys toxicity, but others, called lytic phage, kill *Vibrio cholerae*, so controlling their numbers. Thus numbers of lytic phage and *Vibrio cholerae* are usually inversely related—the more lytic phage the less *Vibrio cholerae* and vice versa. Cholera epidemics in the Ganges Delta generally occur after the monsoons and spring rains, and local scientists have proposed that these cycles are driven by the phage viruses that infect *Vibrio cholerae*.[19] The heavy rains dilute the concentration of phages so allowing more toxigenic and non-toxigenic *Vibrio cholerae* to survive and grow. But when this mixture is ingested by humans only toxigenic *Vibrio cholerae* can survive and multiply in the gut, so those escaping back into the environment via the copious watery diarrhoea will mostly be toxigenic *Vibrio cholerae*. But while this increase in toxigenic vibrios fuels the epidemic it also stimulates a population explosion among the phage that feed on them, and this acts to limit the epidemic by killing off toxigenic *Vibrio cholerae*. This balanced cycle had probably been enacted in the Ganges Delta for many centuries before human travellers transported the microbe around the world.

Although increased standards of hygiene have cleared *Vibrio cholerae* from the developed world, the microbe is still a serious world health problem. When the seventh pandemic hit South America in 1991, *Vibrio cholerae* easily spread among the large susceptible populations in shanty towns and cities with poor sanitation, infecting 400,000 and killing 4,000. And when nearly a million Rwandan refugees fled from tribal conflict to neighbouring Zaire in 1994, the microbe exploited the poor sanitary conditions in the refugee camps of Goma. Twelve thousand died of cholera in a three-week period, a death toll of an alarming 48 per cent.[20] With around 1.5 billion people in the world denied clean drinking water and this number predicted to rise to 3.5 billion by 2025,[21] things could get much worse. Fortunately new oral vaccines are in the pipeline which should, if affordable, be effective in developing countries, working alongside public health programmes to break the cholera cycle in endemic areas.

In this chapter we have seen infectious disease microbes exploiting international travel routes to infect naïve populations worldwide. Many, like the acute childhood infections, have established a global distribution, while others, like plague, yellow fever and cholera are hiding in the environment, waiting for their next opportunity to strike. In the following chapter we look at the devastating impact environmental microbes can have on our lives even if they do not directly infect us.

6

FAMINE AND DEVASTATION

From the beginning of the farming era until relatively recently the majority of the world's population ate a predominantly vegetarian diet. Staples like rice, corn, potatoes and beans provided most people's daily calories, with animal products being luxuries mainly reserved for the rich. So the regular famines which plagued our ancestors were generally caused by crop failures, with adverse weather conditions being the usual culprit. Tropical and subtropical regions were (and still are) the most vulnerable since successful harvests there critically depend on annual rain cycles. In the Indian subcontinent, where searing droughts and devastating floods are commonplace, the anxiously awaited monsoon rains that roll in from the Indian Ocean in early summer are an essential beginning to the growing season. Until recently if the monsoon failed then crops also failed, and thousands, maybe millions, would starve. And hunger is often the trigger for social unrest, with large numbers of malnourished, sick and dying people leaving their homes in search of food and water. Then, as we witness all too frequently in Africa today, where the situation is exacerbated by civil war, these refugees are often housed in appallingly over-crowded camps and feeding stations where sanitary conditions are at best basic, but commonly non-existent. This scenario is the cue for microbes that thrive in such unhygienic conditions to take centre stage, and outbreaks of cholera, typhus and dysentery are almost inevitable.

Microbes that infect domestic crops, and to a lesser extent live-stock, can also precipitate famines. Intensively farmed animals and monoculture crops in particular are an open invitation to microbes, which can sweep through with devastating efficiency and kill the lot. And although in recent times people in the West have felt protected from the effects of famines by the vast international trade in farm produce, this chapter looks at a devastating famine which occurred within the last 150 years in the affluent British Isles.

Several sixteenth-century English adventurers travelling the seas in the reign of Elizabeth I, including Sir Walter Raleigh, Sir John Hawkins and Sir Francies Drake, have been credited with bringing the potato to Britain directly from South America. But in fact it seems more likely that the tuber reached Britain from Europe some time in the 1590s. The Conquistadores, intent on plundering the New World of gold, silver and gems, would not have given the humble potato a second glance, but nevertheless they probably inadvertently introduced it to Spain around 1570 by using it to sustain their ships' crews on the return journey from the Americas.

At first Europeans spurned the potato as 'the food of the devil', but its popularity soon grew, and by the middle of the nineteenth century it was cultivated throughout Northern Europe and the United States, providing a calorific staple that improved the health of the poor enormously. Sir Walter Raleigh, although denied the credit of importing the potato to Britain, may well have been the first to introduce it to Ireland. As the story goes, he gave some tubers to the gardener at Youghal, his estate in county Cork, who planted them, but, not knowing what they were, he ate the (poisonous) berries rather than the root tubers.[1] The word 'potato' comes from *papa* or *patata*, names used by the Indians in the Andean region of South America who have cultivated the plant for thousands of years. The potato we grow today, *Solanum tuberosum*, has been traced to its origin from the wild *solanum* species that grow in the coastal regions of Chile, but by the time Europeans arrived in South America it was grown as a food plant throughout the Andes, and was a mainstay for the Incas in Peru. However, it was unknown in Central and North America until brought there by white settlers in the seventeenth and eighteenth centuries.

The Irish Situation

The potato plant is tolerant of most climatic and soil conditions and this is perhaps why it was particularly welcomed in the wet and misty reaches of Western Ireland, where damp summers and peaty, poorly drained soil often prevented grain crops from ripening. Famine, hunger and death from starvation were all too common in Ireland and the potato seemed to provide a valuable staple which, acre for acre, could feed twice as many as the grains it soon replaced. And uniquely among vegetables, a diet of potatoes supplemented with a little milk provides all the essential vitamins required to stave off malnutrition, so the scurvy that was previously common among Irish schoolchildren became a rarity.[2] All through the eighteenth century, when the population of Ireland was growing rapidly, the importance of the potato in the Irish diet rose until by the early 1800s the poor ate nothing but potatoes from one month's end to the next. Save for the 'meal months' of July and August when the stored potatoes had either been eaten or were rotting and the new were not yet ready to dig, each person consumed an average of 8 pounds of potatoes a day. Even nowadays potatoes cannot be stored for as long as grain, so communities who depend on potatoes as a staple are more liable to famine. Over the years new varieties of potato with improved productivity and storing capacity have been developed, and the Irish Apple, introduced in 1770 could be reliably stored for at least a year. But the Apple was superseded by the Lumper, which arrived from Scotland in 1808. Whereas this was reliable and prolific even in poor soil, it was coarse and tasteless and its storing capacity was inferior to the Apple. The English and Scots scorned Lumpers as only fit for animal feed, but the Irish poor, struggling against hunger, welcomed it and soon it was grown all over Western Ireland to feed the growing population.

By 1845 when the blight first struck the Irish potato crop the lot of the poor labourers and cottiers had reached crisis point. Their conditions had remained virtually unaltered since feudal times; they owned no land, their living conditions were squalid and, encouraged by the Catholic Church, their families were increasing in size. Indeed between 1800 and 1845 the population of Ireland rose from 4.5 million to over 8 million, with around 650,000 poor, uneducated, dependent labourers.[3] They lived and worked on large estates owned by the gentry, who generally only visited to host hunting and shooting parties. These absentee land-lords financed their exploits by letting out land to tenant farmers who, in turn, earned the rent by subletting quarter-acre portions, known as 'conacre' (derived from the words corn and acre), to landless labourers and cottiers. These unfortunates, who lived in thatched hovels without windows or chimneys, where turf fires smoldered incessantly, were at the bottom of a chain of land renters and so proportionally paid the most for it. They scraped the annual rent together by casual labouring on local farms and estates, as well as fattening a pig on potatoes. Each family had to grow enough potatoes on their quarter-acre conacer to feed themselves and the pig throughout the year—an average of 32 pounds a day.

An article in *The Times* newspaper by Mr Thomas Campbell Foster, who visited Ireland in 1845, outlined the annual budget of a typical labourer's family:[4]

Income: £3 18s. from wages (6p a day), £4 0s. from sale of pig
Annual rent: £5 0s. (£2 10s. for cottage, £2 10s. for conacer)
Balance left to spend on clothes, candles, meal, drink, tools etc: £2 18s.

It is plain that if the potato crop failed the family starved.

To the Irish labourer their conacer was their lifeline and they would not give it up at any price. The potatoes grown on it represented not only capital, wages, rent and subsistence but were also used as a form of social currency to finalize land tenancy issues, and in marriage settlements.[5] Landlords complained that their cottiers were lazy, that they resisted all attempts to better themselves and refused to diversify their crops, concluding that they only had themselves to blame when starvation set in. But in fact the poor were caught in a downward spiral—their rent was too heavy and their conacre too small for manoeuvre. In short, they could *not* afford *not* to grow potatoes if they were to feed their growing families. In no other country did the dependence on potatoes reach such heights; even just across the Irish Channel, the English and Scots labourers who were also poor at least ate oats and bread as well as potatoes. So when the potato blight hit Britain in 1845 the situation for the poor was bleak, but uniquely for the Irish cottiers it spelt disaster.

The Potato Blight

Late blight in potatoes is caused by the mould *Phytophthora infestans*, which ranks among the most devastating of plant pathogens. Probably originating along with the potato itself in South

America, its primary target is the potato, but it also attacks other members of the *Solanaceaeum* family, including tomatoes. The mould thrives in a cool damp climate and its microscopic spores (sporangia) drift on air currents and are dispersed from plant to plant by the wind and rain. When sporangia land on a suitably damp leaf they either germinate directly or release zoospores which can swim in the thin film of water. The tiny branching filaments, or hyphae, produced by germinating spores penetrate leaves and tubers, winding their way between the cells to steal nutrients and cause decay. Leaves develop telltale black spots and then wilt and die, but before the whole plant withers hyphae bearing new spores emerge from leaf pores, ready to be carried away by the wind to infect other plants. Rain washes the spores from the leaves to the ground, where they infect and rot the tubers and sit out the winter ready to infect next year's crop.

P. infestans began its journey across the world in the early nine-teenth century, following the potato in its travels and reaching Europe from America in the early 1840s. First seen in Belgium, it spread to Britain by 1845, reaching Ireland in September of that year. The Irish were used to their potato plants developing diseases like 'curl' and 'scab', and knew them to be a sure sign of shortage, hunger and maybe even death to come. But the potato blight was on a different scale entirely. It appeared suddenly and wiped out the whole crop, leaving nothing but a rotting mess. The editor of *The Gardeners' Chronicle*, Professor John Lindley, a distinguished botanist from University College London, first de-scribed the disease in 1845:[6]

...the disease consists of a gradual decay of the leaves and stems, which become a putrid mass, and the tubers are affected by degrees in a similar way. The first obvious sign is the appearance on the edge of the leaf of a black spot which gradually spreads; the gangrene then attacks the haulms (stems), and in a few days the latter are decayed, emitting a peculiar and rather offensive odour.

That smell was to haunt the cottiers of Ireland for the next three years.

In 1845 the blight claimed 40 per cent of the Irish potato harvest, a loss that could perhaps have been sustained without many deaths if it had not returned the next year killing 90 per cent of the crop. Then in 1847 the blight receded, but by that time most cottiers had in desperation eaten their seed potatoes so the crop was much reduced. The blight returned again in 1848 to devastate the plants once more.[7]

First and foremost the potato blight brought hardship to the cottiers at the bottom of the social pyramid, but eventually it was to affect all levels of Irish society. Cottiers grew nothing but potatoes so the equation was simple—when the crop failed they could neither eat nor pay the rent. And even if they survived until the following spring most had no tubers left to plant. Smallholders who also grew corn and perhaps kept a few cows were little better off. If they ate their produce they would be evicted for not paying the rent, and if they paid the rent they starved.

The knock-on effect of this tragedy was felt in every walk of life: with little in the way of rent coming in from the estate, landlords promptly laid off their workers and servants to save on wages. The country was hit by a wave of unemployment; since few could afford to buy food let alone luxuries, shopkeepers, manufacturers, wholesale merchants and craftsmen all went out of business. In the coldest winter in living memory starving families huddled in their cottages and waited to die.

In December 1846 a County Cork Justice wrote to the Duke of Wellington recounting his visit to a coastal hamlet:[8]

In the first (hovel) six famished and ghastly skeletons, to all appearances dead, were huddled in a corner on some filthy straw, their sole covering what seemed a ragged horsecloth, and their wretched legs hanging about, naked above the knees, I approached in horror, and found by a low moaning that they were alive, they were in fever—four children, a woman, and what had once been a man. It is impossible to go through the details, suffice it to say, that in a few minutes I was surrounded by at least 200 phantoms, such frightful spectres as no words can describe.

Thousands of destitute cottiers and smallholders left their hovels, many evicted because of rent arrears, and walked to the nearest town carrying their possessions in an attempt to reach the workhouse before starvation took hold. As one eyewitness, Elizabeth Smith, wrote in her diary,[9] 'God help the people; the roads are beset with tattered skeletons that give one a shudder to look at, for how can we feed or clothe so many.' The workhouses were woefully ill equipped to deal with the scale of the crisis and if they were full there was nothing to do but wait outside for an inmate to die.

Some landlords could hardly wait for the cottiers to abandon their filthy homes before demolishing them and repossessing the land, but of course there were good landlords as well as bad. Lord Kildare of Donegal for example cancelled his tenants' rents for the year 1845, and others helped their tenants to emigrate.[10]

In a country where the people were famed for their friendliness and hospitality, thieving was now rife. Families lucky enough to own a healthy crop of potatoes had to protect it against thieves who came by night and dug them with their bare hands or speared tubers from barns with long poles. Anything that would stave off hunger and starvation became a meal, including dogs, foxes, rats, snails, frogs, hedgehogs, crows, seagulls, seaweed and limpets.[11]

Many impoverished Irish labouring families, seeing no future in Ireland, opted for emigration rather than death from starvation or infection. In fact potato famines and emigration to North America, Canada and Australia, had been ongoing for several years before the great famine of 1845–8, but during those years the numbers escalated. In 1846 120,000 left and nearly twice that number followed in 1847, most settling in the US or Canada. The year 1847 saw 90,000 set out for Canada, but their state of health was so poor that 2,000 died before leaving Ireland and a further 13,000 died in transit.[12]

The results of the three-year famine were horrific: 4½ million people faced starvation, and as they slowly weakened opportunistic infectious diseases moved in to feast off their emaciated bodies. Over a million Irish peasants died and a further 1.3 million were banished overseas. Sir William Wilde, travelling in the West of Ireland in 1850, reported that: 'we pass over miles of country without meeting the face of a human being, and rarely saw animals either'. But he did happen upon 'the smoking ruins of a recently unroofed village, with the late miserable inmates huddled together and burrowing for shelter among the crushed rafters of their cabins'.[13]

By the Act of Union of 1801 Ireland was part of the United Kingdom, so cries for help were directed to London and the government of Sir Robert Peel. But the government was reluctant to provide for the starving Irish. This was partly because they were at first unable to grasp the size of the problem—there had been so many previous food shortages in Ireland that this was initially assumed to be just another on the same scale. But also in class-ridden Victorian society the upper echelons preferred to believe that the poor, and particularly the Irish poor, were dirty, lazy, immoral and rebellious, and brought their problems on themselves. Indeed the belief that their afflictions were sent by God as a punishment is clear in Queen Victoria's speech declaring a day of prayer for the suffering Irish, and ordering the people to ask God: 'for the removal of those heavenly judgments which our manifold sins and provocations have most justly deserved'.[14] Open-handed charity was considered to invite slothfulness and corruption, so Charles Trevelyan, Director of Britain's Relief Programme for the Irish advised that 'relief should be made so unattractive as to furnish no motive to ask for it, except in the absence of every other means of subsistence'.[15]

When the potato blight hit, Peel was in a dilemma over the Corn Law which protected home-producers by imposing a heavy duty on imported grain. This favoured rich landowners on whose support he relied, but on the other hand repealing the law would open up the market for imports from the colonies, empower the merchant class and place Britain at the hub of international trading. So when his Commission of Enquiry reported the extent of the problem in Ireland it gave Peel the excuse he needed to repeal the law. Then cheap maize meal was imported from America and distributed to the starving Irish through Relief Centres. But nothing was provided for free: people had to work for their meal, and road making was the allotted task. So in 1847, 734,000 men, who together with their families represented some 3 million people, constructed roads,[16] and to this day the West of Ireland has a network of roads which, some say, lead from nowhere to nowhere.

While all this human misery was being played out in Ireland there were plenty of suggestions as to the cause of the blight. At the time most believed that miasmic vapours emanating from decaying matter had produced the problem, although some took a different approach and blamed static electricity, perhaps emitted by the recently invented steam trains.[17] But in the pages of *The Gardeners' Chronicle* a battle was being waged between two botanical giants.

Professor Lindley of University College London was convinced that adverse weather conditions were to blame. The summer of 1845 was hot and dry until July when there followed a long gloomy period of cold, rain and fog. According to his theory potato plants grew fast in the early fine weather but then absorbed excess water during the wet period. Without sun for weeks on end plants could not rid themselves of this water by transpiration and so became waterlogged and fatal wet putrefaction set in. A plausible theory perhaps, but the Reverend Miles J. Berkley disagreed. From his parish near King's Cliffe in Northamptonshire, he had gained a reputation among naturalists as an expert on fungi, and once he saw the delicate fringe of mould growing on the leaves and tubers of blighted potato plants he was sure that it was the cause of the problem. He, along with a small band of amateur botanists, including Charles Morren in Belgium and Camille Montagne in France, formulated the 'fungal theory', invoking the mould as the culprit. This revolutionary idea, preceding the 'germ theory' by more than twenty years (see Chapter 8), was not surprisingly met with disbelief. In *The Gardeners' Chronicle* Lindley agreed that moulds and mildews commonly grew on rotting leaves and tubers, but argued that they were merely saprophytes living off the rotting material, not the cause of the rot itself.

No one knew exactly how fungi came to grow on dead matter, and in an age when many still believed in spontaneous generation, some thought that a fungus was a product of the diseased plant itself. Perhaps it was an outward sign of internal disease like the rash of measles or smallpox, or an attempt by a diseased plant to generate a tiny new healthy plant. But the botanists stuck to their guns and in 1846 Berkley published a paper in the *Journal of the Horticultural Society of London*, entitled 'Observations, Botanical and Physiological, on the Potato Murrain',[18] in which he argued that the fungus was not a *saprophyte* but a *parasite* as its hyphae actually invaded the interior of the potato leaf. Montagne, a retired surgeon from Napoleon's army who had first named the mould *Botrytis infestans*, provided drawings of it growing inside an affected leaf (Figure 6.1). Berkley reminded his readers that two diseases of wheat, 'bunt' and 'rust', although nothing like as devastating as the blight, were nevertheless generally accepted as fungal in origin. But in the end he could not demonstrate conclusively that the mould actually caused healthy plants to come down with blight, so his conviction remained just a theory until 1861 when a German scientist, Anton de Bary, eventually unravelled the life cycle of the mould, so proving the 'fungal theory' correct. In 1876 he renamed the mould *P. infestans*, and it has now been reclassified as a heterokont rather than a fungus, more akin to algae and water moulds than mushrooms and toadstools.

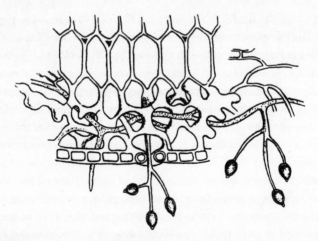

Figure 6.1 A section of a leaf showing the potato blight fungus growing and producing spores (Berkley 1846)

None of this academic wrangling was of any practical use to the starving Irish, but there was no shortage of advice for them on how to prevent or cure the blight. Remedies for saving unblighted potatoes included desiccating; burying in lime; coating in salt; and soaking in chlorine, oil of vitriol, manganese dioxide or copper solution.[19] And although several of these chemicals would have killed much more rapidly than starvation, one useful recommendation was to cut the stems of the plants as soon as the blight appeared, in the hope of saving the tubers below. Interestingly, the Cambrian newspaper reported the chance observation that potato plants growing near a copper-smelting works in Swansea remained healthy while all those around were dying of blight.[20] It is a pity that this was not investigated further since the first fungicide used to treat the blight and other related moulds was Bordeaux mixture, which contained copper salts as its main ingredient.

To start with many cottiers refused to eat the unaffected part of partially blighted potatoes in the belief that they caused diseases like cholera. So to prove this groundless, the enterprising, but perhaps rather foolhardy, Monsieur Bonjean from Chambery in France, spent three days eating 8 pounds of partially blighted potatoes a day and drinking the stinking water in which they had been boiled.[21] Fortunately he survived to tell the tale.

Typhus

Inevitably the starving Irish were prey to all sorts of opportunistic microbes, and countless numbers died from infections before starvation could claim them. The poor flocked to workhouses in their thousands in the hope of finding food and shelter, but these overcrowded establishments, most of which lacked safe drinking water and sewage disposal, probably killed as many as they saved. In these conditions microbes spread by the faecal oral route flourish, and not surprisingly diseases like dysentery, typhoid and cholera were rife. But probably the commonest cause of death, at least for inmates of the workhouses, was typhus, caused by a microbe that exploits poverty and overcrowding to the full. Indeed, the various names used for typhus: 'camp fever', 'jail fever', 'ship fever' and 'hunger fever', graphically depict its haunts and habits.

The name 'typhus' derives from the Greek *typhos*, meaning 'haze' or 'smoke', describing the detached mental state that sufferers experience as the microbe invades their brain causing fatal encephalitis. The typhus microbe is a *Rickettsia*, a type of bacterium that lacks the essentials for free living and so parasitizes cells of other organisms. *Rickettsia prowazekii,* the cause of typhus, is named after two early twentieth-century scientists, both of whom accidentally picked up a fatal dose of typhus while trying to identify its cause. Howard Taylor Ricketts was an American microbiologist who identified the first *Rickettsia* while working in Mexico in 1910, but died before he could show that it caused typhus. Then in 1914 the Bohemian bacteriologist, Stanislaus von Prowazeki, confirmed Ricketts' finding but also died before completing the work. So it was left to the Brazilian bacteriologist, Henrique da Rocha Lima, to finally link the *Rickettsia* to typhus in 1916.

Although today *R. prowazekii* is entirely a human parasite, its DNA sequence suggests that it evolved from its relative, *R. typhi*, an ancient parasite of rats that is harmlessly transmitted by their fleas.[22] Its foray into humans probably began during the farming era when rats first colonized people's homes. At this stage rat fleas must have occasionally bitten humans, sometimes passing on *R. typhi* at the same time, and as living conditions became more and more crowded and squalid, these encounters would have increased until eventually

R. typhi evolved into the human parasite, *R. prowazekii*, which was picked up and transmitted between humans by body lice.

Around the same time as Ricketts and Prowazeki were identifying *Rickettsia*, Charles Nicolle, working at the Pasteur Institute in Tunis, noticed that staff who handled the clothes of typhus patients arriving in hospital often caught the infection, but once patients were on the wards the risk of contagion seemed to disappear. From this he surmised that the microbe was spread by something attached to clothes that could be washed away. He correctly pointed the finger at the body louse, and this astute piece of detective work won him a Nobel Prize in 1928. These tiny crab-like creatures thrive in unhygienic conditions, living off the blood of the poor and dispossessed.

We probably inherited lice (*Phthiraptera*) from our hirsute primate ancestors, but as their hair receded their lice adapted by evolving into three separate types. Some grew claws to grip on to fine hairs and evolved into head lice, while pubic lice became adept at hanging on to coarser pubic hairs. In contrast the body louse abandoned ship altogether, preferring to live among the folds and seams of clothes and bedding. These bloodsucking parasites regularly emerge from their lair to feed off their victims, and on occasions they ingest *R. prowazekii* along with their blood meal. The microbe then multiplies in the louse's gut, eventually rupturing the lining and spilling the bloody contents into its tissues. At this stage the louse turns red, so finding red lice on a sick patient is a good indication that they are suffering from typhus. *R. prowazekii* kills lice in eight to twelve days, but given the right conditions they still have plenty of time to spread the microbe around.

Unlike mosquito or flea vectors, lice do not inject their lethal cargo into their victims while feeding, but leave a trail of microbe-laden faeces on their skin. And since bites cause intense itching, people invariably scratch, and in so doing damage their skin sufficiently to let *R. prowazekii* in. The microbe then colonizes the tissues, particularly targeting the lining of blood vessels, and after fourteen days this produces high fever, and severe head, muscle and joint aches. Damaged small blood vessels may become blocked, causing gangrene of fingers and toes, while bleeding into the skin produces a characteristic dark rash. The microbe invades the brain in up to 80 per cent of cases, causing delirium, seizures and stupor, often progressing to coma and death. Without antibiotic treatment *R. prowazekii* kills up to 60 per cent of its victims.

Being transported from host to host by a louse may not seem a very efficient way to get about, but *R. prowazekii* has certainly caused huge epidemics over many centuries. Those who survive an attack of typhus often carry the microbe for many years, suffering recurrent bouts of a milder disease (called Brill-Zinsser disease after the two doctors who described it) and also acting as a microbe reservoir. Typhus carriers can spark off epidemics wherever and whenever infected lice can spread the microbe. And their job is facilitated by the fact that lice dislike heat, so when victims develop a fever their lice seek out new hosts, carrying the deadly microbe with them. So it is not surprising that during the Irish potato famine when people were destitute, huddling together for warmth, unable to wash, and sharing clothes and bedding, *R. prowazekii* had a field day.

There have been many episodes in our history when *R. prowazekii* got the upper hand. Indeed, in virtually all military campaigns until the Second World War deaths from this and other microbes far outweighed those from combat, a situation only reversed in the mid-twentieth century by a combination of insecticides and antibiotics. Typhus repeatedly dogged the French army throughout the Napoleonic wars, often decimating the troops.[23] When he set out for Russia in the summer of 1812, Napoleon Bonaparte had well over 500,000 soldiers, but his army soon outstripped its supply lines and with little water to drink none could be spared for washing. Typhus struck by the time they reached Poland and as deaths from disease and starvation rocketed, many deserted. When he entered Russia Napoleon had only 130,000 troops left, and by the time they reached Moscow only 90,000 remained to see the smoking ruin of the abandoned city. The story of the retreat through a particularly severe winter is one of unimaginable hardship, and as the remaining men straggled back *R. prowazekii* continued to take its toll along with dysentery, pneumonia, starvation and frostbite; only 35,000 reached home alive. Undaunted, Napoleon raised another 500,000-strong army in 1813 to fight the Germans, but again a typhus epidemic among the troops played a key role in his defeat at the Battle of Nations. Many historians believe that it was primarily *R. prowazekii* that caused Napoleon 1's dream of European domination to fail.

As both general and personal hygiene improved *R. prowazekii* found it increasingly difficult to hold its own, and by the mid-1880s typhus was rare in the West. But the microbe retained a foothold in the East. During the First World War it killed thousands fighting on the Eastern Front and in the years that followed, an epidemic in Russia that killed around 3 million prompted Lenin to say 'Either socialism will defeat the louse or the louse will defeat socialism'.[24]

Most of the 1.3 million driven from Ireland by the threat of starvation travelled from Liverpool to New York where, as you can imagine, boatloads of starving Irish were not always welcomed. Americans genuinely feared the epidemics the immigrants might spawn and for a time the Irish were made scapegoats for any outbreak that hit the city. But in fact it was the conditions the Irish were forced to endure in Ireland, on board ship, and then in their new homes that were the culprit because they encouraged microbes to thrive. New York lagged behind most large European cities in sanitary reform,[25] so often the destitute, malnourished and penniless Irish arriving in the city just substituted their poorly ventilated, overcrowded, Irish hovels for the cramped attics and basements of New York tenements. With no running water or sewage disposal it is no surprise that Irish immigrants were attacked by all the microbes that find a niche in these appalling living conditions—typhus, typhoid, cholera, dysentery, tuberculosis, and many more.

Typhoid

Typhoid, or enteric fever, is a form of food poisoning caused by *Salmonella (S.) typhi*. These days epidemics classically follow a meal of shellfish grown in water contaminated with sewage, but the microbe can also be spread by unwashed hands. *S. typhi* causes many mild or even silent infections, but in those who come down with classic typhoid fever the microbe penetrates the gut lining, multiplies in the blood and invades the organs. The first signs of anything amiss appear two weeks later with a stepwise rising fever followed by the characteristic rash of rose-coloured spots. The microbe then homes back to the gut where it erodes the lining, producing deep ulcers which sometimes perforate the gut wall to cause fatal bleeding or peritonitis. The death rate in untreated cases is around 10–20 per cent.

After recovery *S. typhi* often hides in the gall bladder from where it has access to the gut and is excreted in faeces, so apparently healthy carriers can spread the microbe. The first identified typhoid carrier in the US was Mary Mallon, now famously dubbed 'Typhoid Mary'. She was born in 1869 in Cookstown, Ireland, emigrating to New York when she was fifteen years old. Once there she entered domestic service and developed a flair for cooking. Over the years she was employed by several wealthy New Yorkers, and during the summer of 1906 she worked as cook for the family of a New York banker, Charles Henry Warren, who had rented a summer house at Oyster Bay, Long Island. When six of the eleven inhabitants were struck down with typhoid the owner of the house hired George Soper, a sanitary engineering expert, to investigate. Having excluded the common causes of milk, water, bay clams and other foods, he pointed the finger at the new cook. By tracing her recent moves he uncovered twenty-two cases of typhoid with one death in seven of the households she had previously worked in; all had experienced a typhoid outbreak within a few weeks of her arrival. On this evidence Mary was apprehended, and although she insisted that she had never suffered from typhoid, she was obviously a chronic carrier since her faeces were teaming with *S. typhi*. The poor woman was detained on North Brother Island in the East River, housed in a cottage in the grounds of Riverside Hospital. She campaigned tirelessly for her release and in 1910 the Health Board agreed to let her go on the understanding that she would not work as a cook again. But having given her word, she was found in 1915 masquerading as Mrs Brown, employed as a cook in the Sloane Maternity Hospital in Manhattan where a typhoid outbreak infected twenty-five people, killing two. After that Mary spent the remaining twenty-three years of her life imprisoned on North Brother Island, working in the hospital laboratory.

Tuberculosis

Given the living conditions the Irish poor had to contend with both before and after emigration to the US, it is not surprising that in cities like New York they suffered disproportionately from tuberculosis (TB)—so did poor Blacks and Jews. This disease has been recognized for several millennia, with evidence of its ravages found in human remains from both the Old and New Worlds, including Egyptian mummies and skeletons from early Hindu, Greek, Roman and pre-Colombian American archaeological sites.[26] Until recently scientists thought that *Mycobacterium (M) tuberculosis,* the bacterium that causes human TB, evolved from *M. bovis,* the TB microbe of cattle. They estimated that the microbe jumped from cattle to humans some 15,000–20,000 years ago, but recent molecular evidence refutes this claim. Although we still do not know the precise date at which the different strains of mycobacteria evolved from their common ancestor, molecular clock analysis of human strains suggests that they arose in Africa well before the evolution of *M. bovis* and so could not have arisen from it.[27]

This evolutionary scenario, if correct, places *M. tuberculosis* among the very ancient human pathogens. However there is no doubt that the microbe, which spreads best in crowded, unhygienic, poorly ventilated surroundings, rose to prominence as towns and cities expanded, reaching a peak in the US and Europe in the early 1800s when almost everybody in large cities like London, Paris and New York was infected, and it was among the commonest causes of death.

M. tuberculosis is spread by airborne droplets of infected sputum coughed, sneezed or otherwise exhaled by sufferers. Transmission is generally restricted to household contacts where typically young children pick up the microbe from elderly relatives. *M. tuberculosis* requires a high concentration of oxygen to survive so once inhaled bacteria head for the apex of the lungs where they set up a lifelong focus of infection. They are well equipped to fight off the body's immune attack and indeed take up residence inside the macrophages sent to the area to destroy them. In healthy people the infection is controlled by the immune system, but if the body's immunity is depressed by disease, old age or malnutrition it is likely to reactivate.

M. tuberculosis can attack almost anywhere in the body: TB of the spine, called Pott's disease, leads to painful deformity and eventual paralysis, while lymph gland swellings known as 'scrofula' may rupture through the skin, producing a continuous flow of pus. But the symptoms of active TB are most often caused by the bacteria slowly but relentlessly destroying the lungs. This terrible disease, which was called 'consumption' because of the extreme wasting it causes, killed many famous people in the nineteenth century, including literary giants such as Emily and Anne Bronte, the poet John Keats, and the playwright Anton Chekov, and has been romanticized in Verdi's *La Traviata* and Puccini's *La Bohème*. Indeed in the nineteenth century the pale, drawn appearance of the consumptive was considered so attractive that the poet Lord Byron is said to have remarked 'I should like to die of a consumption, because the ladies would all say, "Look at poor Byron. How interesting he looks dying"'.[28]

But in reality the disease is far from romantic. While the bacteria eat away at the lungs, causing sufferers to struggle for breath, their emaciated bodies are wracked by bouts of fever, drenching sweats and spasms of coughing, leaving them completely exhausted—undoubtedly an unenviable slow and painful death. Even with the effective TB treatment developed in the twentieth century the microbe's stranglehold on the poor in big cities remained, and, as we will see in Chapter 8, is as much of a worldwide problem now as it was in the nineteenth century.

The potato blight fungus, by doing what it is programmed to do—infect and reproduce in potato plants—precipitated one of the worst famines in recent times. By wiping out Ireland's staple food crop the fungus set in motion a chain of events that saw the starving Irish prey to dozens of opportunistic microbes, all flourishing on their emaciated bodies. The result was a million dead, 1.3 million emigrated and Ireland depopulated, never to be the same again. Fortunately the potato never regained its prominent position in the Irish diet and the potato blight fungus, although still around, has thus been denied the opportunity to prosper on quite such a grand scale.

The consequences of the blight were far-reaching for Ireland, Britain and the countries in which the Irish made their new homes. In Britain the repeal of the Corn Laws was the beginning of free trade, enhancing the country's position at the centre of the commercial world, while in the US the Catholic faith that arrived along with the Irish heralded an era of religious as well as racial conflict. Now, 150 years later, the descendants of those Irish immigrants number some 34 million and are influential in all walks of life. Among the most prominent were two American Presidents, John F. Kennedy and Ronald Reagan.

7

DEADLY COMPANIONS REVEALED

For all but the last 150 years our ancestors survived the on-slaught of infectious diseases without the least understanding of their cause, and virtually no effective treatments. Although through the ages many theories emerged to explain these phe-nomena, they were generally misguided, and the treatments they invoked usually did more harm than good. In fact right up until the eighteenth century most herbal remedies used by doctors, although they may on occasions have relieved suffering, con-tained no active ingredients; the best advice a doctor could offer during epidemics was to flee or pray (or both).

Our present knowledge of microbes and the diseases they cause was mostly accumulated painstakingly slowly by observing and documenting cases and experimenting with remedies. This labori-ous process was interspersed with a few 'eureka moments' when a major discovery suddenly changed our way of thinking and opened up whole new avenues to be explored. Now that we understand many of the steps leading from microbe to epidemic we have begun to use this knowledge to fuel the fight against microbes. But unaware of their invisible cause, our ancestors not surprisingly attributed epidemics to supernatural forces beyond their control.

In the great civilizations of the past, cults grew up to explain the inexplicable, usually invoking an angry god sending epidemics as a punishment for wrongdoing. In Egyptian mythology, for example, the goddess Sekhmet, depicted with a lion's head and woman's body, caused pestilence when angry, and had to be appeased with offerings and sacrifices. Much later, during the Black Death, the flagellant movement emerged in Germany and eventually spread through most of Europe. This strange religious sect aimed to appease their god by their suffering and induce him to relent and remove the terrible plague. Wearing sackcloth and ashes, the brotherhood marched from town to town, holding services in local churches during which they worked themselves into a state of mass hysteria and lashed their naked bodies until blood streamed from their wounds, and some even died of their injuries.

In this chapter we track our progress towards enlightenment about the cause of infectious diseases, and how they can be prevented and treated. In particular we look at the history of our fight against the smallpox virus, which began as a serial killer and has now been completely eradicated from the wild.

The Greek physician Hippocrates of Cos, living in the fourth century BC, was the first to discard superstitious and religious beliefs, instead attributing diseases to an imbalance of the four humours: blood (sanguine), yellow bile (choleric), black bile (melancholic) and phlegm (phlegmatic). This 'humoral theory' remained influential in Europe until the end of the seventeenth century, but Hippocrates' greatest contribution to medicine, earning him the title 'Father of Modern Medicine', was his detailed descriptions of specific diseases. Prior to this ill health was just ill health, but, by carefully documenting the symptoms of thousands of his patients, Hippocrates distinguished one disease from another and categorized them as epidemic or endemic. This was a huge step forward, but unfortunately his example was not followed by his successors, and although the influential Galen of Pergamum, physician to the Roman Emperor Marcus Aurelius Antonius during the Antonine Plague in AD 166 (see Chapter 3), upheld Hippocratic principles, he believed that epidemics resulted from an imbalance in the atmosphere which could assume an 'epidemic constitution'. And since this theory could not be proved or disproved, it held sway for many centuries; but finally, as new and successful drug treatments were introduced in the sixteenth and seventeenth centuries, people began to question its wisdom.

One such new treatment was for the ague (malaria), which was rife in Europe at the time. In accordance with Galen's teaching it was treated by bleeding and purging to release the humours. But in the 1630s word came from South America that the bark of the cinchona, or fever tree, which grows in the forests of the Andes, could effect a cure. According to Antonio de Calancha, an Augustine monk writing at the time: 'A tree grows which they call "the fever tree" (*arbol de calenturas*) in the country of Loxa, whose bark, of the colour of cinnamon, made into a powder amounting to the weight of two small silver coins and given as a beverage, cures the fevers and tertianas; it has produced miraculous results in Lima.'[1] We now know that the bark contains quinine and several other substances active against malaria.

During the Middle Ages there was widespread belief in the 'miasma theory', which extended the principles of Hippocrates and Galen by attributing epidemics to foul smells and noxious vapours emanating from swamps and rotting organic material. This theory persisted in the West until the nineteenth century, and although misguided, was instrumental in driving a much needed clean up of towns and cities long before microbes were incriminated as the cause of the problem. The threat of the first cholera epidemic in England in 1832 began this process in earnest, and under the leadership of the ardent reformer Edwin Chadwick, new and efficient water and sewage systems were installed to rid cities of their foul-smelling filth. These improvements soon spread to cities in Europe and then the US, and with coincident improvements in housing and health services, at last Western cities were much healthier places to live.

The remarkable Italian doctor Girolamo Fracastoro (1478–1553)— the same gentleman of Verona who named syphilis after a shepherd in his poem on the subject (see Chapter 5)—published a treatise in 1546 proposing that epidemic diseases like smallpox and measles were caused by seminaria (seeds) which spread the contagion from one person to another. He envisaged these seeds spreading by three possible routes: direct contact, contamination of inanimate objects like clothes and blankets, or through the air. Fracastoro's seeds may not quite resemble living bacteria, but they are uncannily like viruses. So he was the first to propose the 'germ theory' which challenged the miasma theory and provided a topic for intellectual debate for the next 300 years. But its proof required visualization of the microscopic world, and that had to wait until the Dutchman Antoni van Leeuwenhoek made a microscope powerful enough to reveal microbes in the seventeenth century. Van Leeuwenhoek's interest in magnifying lenses came from his work in the textile trade, where he used them to count the thread density of cloth. But he also had a lively interest in the natural world and pioneered a new way of grinding powerful lenses to make microscopes, through which he looked at anything from bee stings to spermatozoa. In 1676, while peering at a pepper that he had soaked in water for three weeks (to find out if the spicy taste was caused by prickly spikes), he was surprised to see tiny 'animalculae' moving about in the water under his lens. He described these as 'incredibly small, nay so small, in my sight, that I judge that even if 100 of these very wee animals lay stretched out one against another, they could not reach to the length of a grain of coarse sand'.[2] And to his obvious amazement, once he started looking he found these 'animalculae' everywhere:

The number of these Animals in the scurf of a man's teeth are so many that I believe they exceed the number of Men in a kingdom. For upon the examination of a small parcel of it, no thicker than a Horse-hair, I found too many living Animals therein, that I guess there might have been 1000 in a quantity of matter no bigger than the 1/100 part of a sand.[3]

Van Leeuwenhoek suggested that these 'animalculae' were the cause of infectious diseases, a theory that was only proved correct over 200 years later when the first pathogenic bacterium was isolated. In the meantime there was still a strong belief in spontaneous generation of living things, which was only dispelled when Louis Pasteur, the famous French microbiologist and chemist, conclusively demonstrated that by using filters to exclude dust particles he could prevent mould from growing in boiled broth. This convinced most Europeans that moulds did not grow spontaneously but required to be seeded by 'germs' from the air, and so the germ theory began to take hold.

Then the British surgeon Joseph Lister, hearing of Pasteur's experiments, put two and two together and realized that airborne germs must be responsible for the wound infections which killed around half of those undergoing surgery. Lister pioneered strict antiseptic surgical techniques and in 1871 he invented a carbolic acid spray intended to kill germs in his operating theatre. Using this he virtually overcame the wound sepsis problem in his surgical units at Glasgow and Edinburgh, and thanks to his methods German doctors were able to save the lives of hundreds of soldiers wounded in the Franco–Prussian war. Despite these successes most surgeons in the UK and the US were still opposed to the new ideas. But by the 1880s, after Robert Koch had demonstrated that the use of steam to sterilize surgical instruments also reduced wound sepsis, and Lister had successfully carried out complex surgical procedures under full antiseptic cover, the doubters were gradually convinced and surgery became a much safer procedure.

The golden age of bacteriology began in 1877 when Koch isolated the anthrax bacillus, the cause of the zoonotic disease, anthrax. There followed a spate of microbe discoveries, so that by the end of the nineteenth century those causing diphtheria, typhoid, leprosy, pneumonia, gonorrhoea, plague, tetanus and syphilis had all been identified. Koch himself isolated the tuberculosis bacillus in 1882 and the cholera vibrio in 1883. He established strict scientific criteria for linking a microbe to a disease, which are now known as Koch's postulates. To prove a causative association the microbe must be:

- found in the body in all cases of the disease;
- isolated from the case and grown and maintained in a pure culture;
- capable of reproducing the disease when the pure culture is inoculated into a susceptible animal;
- retrieved from an inoculated animal.

Those opposed to the germ theory fought it for some time but were soon overcome by the sheer weight of experimental evidence, and it was generally accepted by the beginning of the twentieth century. Koch was awarded a Nobel Prize for his work on tuberculosis in 1905.

But there were still many common infectious diseases such as smallpox, measles, mumps, rubella and flu for which no microbe could be found and their cause remained a mystery. Called 'filterable agents' because the infectivity passed through filters that retained bacteria, at the time most people believed that they would just turn out to be very small bacteria. But with the invention of the electron microscope in 1932, it was obvious that these infections were caused by a very different type of microbe—a virus.

The history of our struggle against the smallpox virus is worth following as a unique journey from ancient beliefs and superstitions to the triumph of complete global eradication. In the towns and cities of our oldest civilizations smallpox was a disease that could not be ignored; it regularly swept through like a giant wave, killing up to a third of those infected and scarring, blinding or otherwise maiming many of the survivors. In many civilizations this much feared killer had its own god, like the Chinese goddess T'ou-Shen Niang-Niang and the Indian folk goddess Shitala, to whom the people prayed and made offerings in the hope of being spared or cured. In AD 450 the Bishop of Rheims, later canonized as St Nicaise, patron saint of smallpox victims, survived a smallpox epidemic brought to France by the marauding Huns, only to be beheaded by these Eastern invaders a year later on the steps of his own cathedral. He can still be seen there today, carved in stone, standing above the north door holding his mitred head in his hands.

Smallpox was first distinguished from measles by the Persian physician Al-Razi (Rhazes, *c.*865–925/932), chief of the hospital in Baghdad, in his 'Treatise on the Smallpox and Measles'. However, Al-Razi's main claim to fame was the 'heat treatment' which he advocated for smallpox and which was faithfully followed by many well-meaning doctors until the seventeenth century, much to the detriment of their unfortunate patients. He and his followers believed that smallpox resulted from fermentation of the blood, and the humours this produced escaped through pores in the skin. In an attempt to assist this process, the 'heat treatment' confined sufferers to a room with a blazing fire and sealed windows to sweat out the offending humours. And throughout Europe from the twelfth century onwards this was supplemented with the 'red treatment', which condemned smallpox victims to be dressed in red clothes, wrapped in red blankets, confined to rooms hung with red drapes and attended only by people dressed in red. This was thought to reduce scarring from the pocks and was religiously applied to all royal smallpox sufferers (see Chapter 4). In the fourteenth century Charles V of France was dressed in red shirt and stockings and wore a red veil: he survived, but nearly 400 years later the great Hapsburg Emperor Joseph I who was swaddled in 20 yards of red cloth was not so fortunate; his death cost Austria the Spanish throne. This bizarre practice of 'erythrotherapy' probably originated in Japan where according to folklore red is the colour for expelling demons and illness, but despite being discredited in the early 1900s, it was not completely abandoned until the 1930s. With the additional application of leeches, a widespread treatment to 'bleed out' a fever, it is not surprising that people who could afford smallpox treatment died more frequently than those who could not.

But all this changed in the mid-seventeenth century when the English physician Sir Thomas Sydenham noticed the increased smallpox mortality among the rich and linked it to the treatment they received. He improved survival by advocating no treatment for milder cases, declaring that they would recover unaided, and introducing a 'cooling treatment' for the more lethal confluent smallpox, with windows flung open to dispel the evil humours.

Smallpox Inoculation

Inoculation against smallpox was going on in both China and India for hundreds of years before it was brought to the West, and since different methods were used, the practice almost certainly arose independently in each of these countries.[4] The technique of inoculation (also called variolation or engrafting) is first mentioned in Chinese medical books dating from AD 1500, but it had probably been practiced there from around AD 1000. According to legend, the technique was introduced by a reclusive nun who lived in a reed hut on the top of the sacred Omei Mountain. She claimed to be the incarnation of the Goddess of Mercy come to preserve the lives of children by implanting smallpox, which she did by preparing a powder from dried 'scabs' and blowing it up the child's nose (left nostril for a girl, right for a boy) through a silver tube. Six days later the child came down with a fever and developed pocks, but most recovered and were then immune from smallpox for life. As the practice spread throughout the region the nun attracted quite a following, and after her death she was worshipped locally as the goddess of smallpox.

The Indian technique used needles dipped in pus taken from smallpox pustules to puncture the skin at several sites on the upper arm or forehead, and the small wounds were then covered with a paste made from boiled rice. As trade routes from India opened up, this practice spread to South-West Asia and into parts of central Europe and Africa, reaching Constantinople towards the end of the seventeenth century. It was here that Lady Mary Wortley Montague 'discovered' inoculation, introducing it to Britain and eventually to the rest of Europe and the US.

Several years before Lady Mary's 'discovery' reports of inoculation reached the Royal Society in London. Indeed the Chinese intranasal method was reported twice in 1700, and then, in 1714 and 1716, papers describing the Turkish practice were delivered by physicians who had seen it for themselves; but these reports were universally ignored.[5] Lady Mary (née Pierrepont, daughter of the first Duke of Kingston) seemed to have everything: intellect, beauty, great riches, and a social standing bestowed by her father's dukedom and her husband's parliamentary position. She must have grown up under the spectre of smallpox as she lived in London during the height of its deadly power, and shortly after her marriage to Edward Wortley Montague MP in 1712 her beloved brother Will died of the disease. Then in 1714, just after the birth of her son Edward, she herself contracted smallpox. And although she recovered, and her baby son remained well, her beautiful face was scarred and her eyelashes gone for ever. This, then, was the situation when Edward Wortley Montague was appointed British ambassador to the Ottoman Empire, and the whole family, plus a large entourage including their personal physician, Dr Charles Maitland, left for Constantinople (now Istanbul) in 1716. With her recent personal experiences of smallpox, it is not surprising that Lady Mary was receptive to any means of protecting her child, so when she heard of the practice of inoculation she was keen to investigate further. After just a few weeks in Constantinople she wrote to her friend Sarah Chiswell:

The smallpox, so fatal, and so general among us, is here entirely harmless, by the invention of *ingrafting*, which is the term they give it. There is a set of old women, who make it their business to perform the operation, every autumn in the month of September, when the great heat is abated.

People send to one another to know if any of their family has a mind to have the small-pox; they make parties for this purpose, and when they are met (commonly 15 or 16 together), the old woman comes with a nut-shell full of the matter of the best sort of small-pox, and asks what vein you please to have opened.

She immediately rips open that you offer her, with a large needle (which gives you no more pain than a common scratch) and puts into the vein, as much of the matter as can lie upon the head of her needle, and after that binds up the little wound with a hollow bit of shell; and in this manner opens up 4 or 5 veins.

The children or young patients play together all the rest of the day, and are in perfect health to the eighth. Then the fever begins to seize them, and they keep their beds two days, very seldom three. They have very rarely above twenty or thirty [pocks] on their faces, which never mark, and in eight days they are as well as before their illness'.[6]

Dr Charles Maitland learnt the procedure from 'an old Greek woman', and in 1718 he inoculated five-year-old Edward on both arms. It was a complete success; after seven to eight days the child developed a fever and around 100 pocks, but they all healed without scarring. The family was back in London again by 1721 when a severe smallpox epidemic broke out and Lady Mary persuaded Dr Maitland to inoculate her four-year-old daughter, Mary. This time two eminent physicians attended as eyewitnesses, one of whom was the influential Sir Hans Sloane, President of the Royal Society and the King's physician. The inoculation was a success, and soon Caroline, Princess of Wales, was keen for her two daughters to be inoculated, but not before the procedure was tested on others. Six condemned criminals from Newgate Prison were bribed with the offer of freedom if their inoculations were successful, and after Maitland had performed the task all six prisoners were released (they were probably already immune to smallpox anyway). When further tests on orphans from the Parish of St James's in London were also uneventful, King George I gave consent for his granddaughters to be inoculated. The success of these high-profile inoculations did much to popularize the procedure in Britain.

But inoculation had its opponents. Many medical practitioners were genuinely concerned about the danger of full-blown smallpox which could result directly from inoculation or by spread of the virus from recently inoculated subjects to the non-immune. However, some were more concerned about their loss of revenue from smallpox cases, and others were just inherently prejudiced against such foreign practices. The words of Dr William Wagstaffe, physician at St Bartholomew's Hospital in London and Fellow of the Royal College of Physicians and of the Royal Society, sum up the attitude common at the time:

Posterity will scarcely be brought to believe that a method practiced only by a few *Ignorant Women*, amongst an illiterate and unthinking People should on a sudden, and upon a slender Experience, so far obtain in one of the most Learned and Polite Nations in the World as to be received into the *Royal Palace*.[7]

Religious arguments were also strongly voiced and were fuelled by the first highly publicized deaths from full-blown smallpox following inoculation. The views expressed by the Reverend Edmund Massey in a sermon preached in 1722 at St Andrew's Church, Holburn, were typical of those arguing against the 'dangerous practice because inoculation opposes the will of God, who sends disease (including smallpox) either to try our faith or to punish us for sins'.[8]

Lady Mary was eloquent in refuting all argument, and many eminent medical men supported her. One such, James Jurin, presented a paper to the Royal Society in 1723, comparing the risk of inoculation with that of natural smallpox and showing that in the recent epidemic one in five or six smallpox victims died, whereas only one out of ninety-one died after inoculation.[9] After this inoculation became an accepted practice in Britain, and successfully protected people from smallpox until it was superseded by a different and safer procedure called vaccination in the early nineteenth century. However, the practice was slower to catch on in Europe, particularly in France where medical prejudices and religious objections held sway for much longer than in Britain.

Smallpox Vaccination

Despite the popularization of inoculation in Britain its uptake was patchy. In rural areas where many villagers reached adulthood without ever being exposed to smallpox, and an epidemic could therefore potentially carry off all the younger generation, inoculation was welcomed with enthusiasm. In contrast, in large cities like London where smallpox was endemic, it was an accepted threat to children from poor families, and no one thought it worth preventing. So smallpox continued to take its toll throughout the eighteenth century.

Edward Jenner (1749–1823) grew up in the town of Berkeley in Gloucestershire, England, and after his medical training at St George's Hospital, London, returned there in 1773 to indulge his two interests of medicine and natural history. Interestingly, he was elected Fellow of the Royal Society not for his work on smallpox but for his discovery that the cuckoo lays her eggs in the nests of other birds, and that once hatched the young parasite takes full advantage of its adoptive parents by ejecting their natural offspring. Be that as it may, his name is immortalized for his later pioneering work on smallpox vaccination, a unique form of preventive medicine that has saved countless millions of lives.

Living in the country, Jenner had heard the traditional tales about cowpox infection protecting people from smallpox, and set out to investigate. Cowpox is a natural skin infection of cattle that causes blisters on cows' udders. It can spread directly to humans and consequently usually affects the hands of milkers. In his practice Jenner often performed inoculations and he saw for himself that 'cowpoxed milkers' did not develop pocks after inoculation, seeming to be already immune. This led him, in 1796, to his famous (and infamous) experiment on young James Phipps, inoculating him with cowpox from a 'pock' on the hand of a milkmaid, Sara Nelmes. The boy developed a mild case of cowpox, but when six weeks later Jenner inoculated him with pus from a natural case of smallpox he remained healthy, indicating that he was immune. In a letter to a friend Jenner wrote:

I was astonished at the close resemblance of the Pustules in some of their stages to the variolous Pustules. But now listen to the most delightful part of my story. The Boy has since been inoculated for the Smallpox which as I ventured to predict produced no effect. I shall pursue my Experiments with redoubled ardor.[10]

And he did. With the next cowpox outbreak in 1798 he inoculated five more children, three of whom he challenged with natural smallpox, and all were immune. In the same year he announced his results in a pamphlet entitled *An inquiry into the Causes and Effects of Variolae Vaccinae, a Disease, Discovered in some of the Western Counties of England, particularly Gloucestershire, and known by the Name of Cow Pox*.

His 'discovery' was soon verified by others and shown to be much safer than inoculation. And Jenner's demonstration that vaccine material could be obtained from pocks of vaccinated children meant that a chain of arm-to-arm vaccinations could be set up without constant recourse to a milkmaid or cow with the pox. With this method of propagation the practice of vaccination rapidly spread throughout the UK, to Europe and then around the world. Indeed in 1803 Charles IV of Spain, who had lost a child of his own to smallpox, sponsored the Balmis-Salvany expedition, named after the doctors who led it, to take vaccination to the Spanish American colonies. They set sail from the Spanish port of La Coruna, with twenty-one orphans from the local orphanage on board to keep the vaccine alive during the voyage by forming a chain of arm-to-arm vaccinations. By 1801 over 100,000 people in the UK had been vaccinated, and its impact was dramatic. Smallpox deaths in London fell from 91.7 per thousand deaths in the late eighteenth century to 51.7 between 1801–25, and to 14.3 between 1851–75. In Sweden where vaccination was widespread, and compulsory from 1816 onwards, the total number of smallpox deaths dropped from 12,000 in 1801 to just 11 in 1822, and during the same period average life expectancy rose from thirty-five to forty years for men and thirty-eight to forty-four for women.[11]

Although there were those who objected to vaccination on religious grounds, believing that it was interfering with the will of God, the relative ease of acceptance of the practice was thanks to the battles fought over its forerunner, inoculation, which it now replaced as a safer option. Jenner's fame was legendary even in his lifetime, and eighty years after his death the French microbiologist Louis Pasteur, in his honour, coined the term 'vaccination' for all inoculations against infectious diseases, saying in his address to an international medical congress in London in 1881:

I have given to the term vaccination an extension which science, I hope, will consecrate as a homage to the merit of and to the immense services rendered by one of the greatest Englishmen, your Jenner. I am indeed happy to be able to praise this immortal name in the noble and hospitable city of London.[12]

The term is still used today.

In his 1801 pamphlet, *The origin of the vaccine inoculation*, Edward Jenner himself suggested that: 'the annihilation of the Small Pox, the most dreadful scourge of the human species, must be the final result of this practice'.[13] He was right, although there were many hurdles to overcome before this could be accomplished.

Arm-to-arm vaccination was cumbersome, often led to supply problems, and was soon found to spread diseases, particularly syphilis. One alarming incident occurred in Rivalta, Italy, in 1814 when sixty-three children were unwittingly vaccinated with material from an apparently healthy infant who was suffering from congenital syphilis. Forty-four of the vaccinees developed syphilis; several died and others passed it on to their mothers and nurses.[14] These problems were eventually overcome when regular supplies of vaccine were obtained from calves inoculated with cowpox at multiple sites on the flank, and this prepared the way for large-scale production of a standardized product. Another problem became apparent when, around twenty years after vaccination was introduced, smallpox reappeared in Europe. Although death rates were low compared with the ravages of the eighteenth century, now the pattern of the epidemics had changed, with adults bearing the brunt of the infection while children (who had been recently vaccinated) were spared. And so it became clear that vaccination did not give lifelong protection, and re-vaccination would be required at intervals throughout life.

By 1896, the 100th anniversary of Jenner's discovery, a successful vaccination strategy had been worked out and global eradication could be contemplated, although the nature of the infectious agent was still unknown.

The World Health Organisation (WHO) announced its World-wide Smallpox Eradiction Campaign in 1966, by which time the virus had already been eliminated from Europe and the US, but was still endemic in thirty-one countries forming four major blocks: South America, Indonesia, sub-Saharan Africa and the Indian subcontinent. The WHO team, headed by Don Henderson, seconded from the Center for Disease Control, Atlanta, USA, argued that, since the virus has no human or animal reservoir, interrupting the chain of infection was key to its worldwide elimination. Their attack was three-pronged: maintain vaccination rates at over 80 per cent, isolate cases to prevent spread, and trace and isolate contacts. This was so successful that they achieved their goal in ten years and finally celebrated a world free of smallpox in 1980, less than 200 years after Jenner set the train in motion by vaccinating James Phipps. A disease that had killed over 300 million in the twentieth century alone was no more.

There is no doubt that the world would be a safer place without the smallpox virus, but instead of completely destroying stocks at the end of the eradication campaign the virus was stored at the US Center for Disease Control and Prevention, Atlanta, and the Research Institute for Viral Preparations in Moscow. These stocks were due to be destroyed at the end of the twentieth century, but when D-Day arrived the decision was postponed by those who suggested that the virus might be a valuable research tool in the future, and by conservationists who could not countenance the deliberate elimination of any 'living' species.

While the argument raged on, the monumental events in the US of 9/11, followed by the anthrax bioterrorist attack in October and November 2001, completely changed the goalposts. Now everyone admits that bioterrorism is a real threat and smallpox is right up there with anthrax and botulinal toxin as one of the most effective weapons. As vaccination ceased shortly after its eradication, most of the world's population is now susceptible to the smallpox virus and stockpiles of vaccine are woefully inadequate. The virus is stable, remaining active for long periods; it is easy to grow up in bulk, and spreads through the air. What more could a terrorist group ask for? And no one really knows how much virus there is out there and whose hands it has fallen into. In the 1980s it was rumoured that Russian military scientists were making even more lethal viruses, such as a smallpox/Ebola hybrid,[15] and with the break-up of the USSR in the 1990s these scientists dispersed to other countries, perhaps taking the virus with them. When the US government decided to vaccinate at-risk groups in preparation for a possible bioterrorist attack, the side-effects of the vaccine, particularly cardiac problems, were unacceptably high compared with the perceived threat. So at the moment those who advocate hanging on to the virus have won the day; research into safer vaccines and antiviral drugs is under way and there is no saying when, if ever, the virus will be completely destroyed.

The incredible success of smallpox vaccination in the eighteenth century showed that prevention of infectious diseases was possible, but it was over eighty years before the next successful vaccines were developed. Again without knowing the nature of the causative microbe, Pasteur developed a vaccine against rabies, and while it was still undergoing tests in animals he was persuaded to use it on a young boy, Joseph Meister, who had been severely savaged by a rabid dog. It saved the boy's life, and this success not only made Pasteur famous throughout Europe but gave a boost to vaccine research and production. Pasteur discovered that bacteria 'weakened' by prolonged growth in the laboratory often proved ideal vaccines as they lost their ability to cause disease but still induced immunity. As more and more microbes responsible for acute infectious diseases were isolated during the late nineteenth and early twentieth centuries, vaccine development became a priority. Now vaccination is recognized as the most cost-effective disease-prevention strategy and virtually all children in industrialized countries are routinely vaccinated against TB, tetanus, diphtheria, pertussis, mumps, rubella, measles and polio, and many of these diseases are fading into memory. Newer vaccines against killers like yellow fever, hepatitis B, pneumococcal pneumonia and rotavirus are now available, and WHO are set on sending diseases such as measles, polio, rabies, leprosy and hepatitis B to follow smallpox into oblivion.

The Discovery of Antibiotics

Before the antibiotic era some infectious diseases could be prevented with vaccines but none could be cured. So when antimicrobial drugs arrived on the scene they seemed like a miracle cure. The sulphonamides, the first drugs with antibacterial activity, were discovered in 1932. After testing hundreds of different compounds, Gerhart Domagk, a medical researcher at the Bayer research laboratories in Germany, found prontocil, a chemical dye, which could cure a variety of hitherto lethal bacterial infections. By 1937 more active derivatives of prontocil, the sulphonamides, were available, and among the microbes sensitive to the drugs was *Streptococcus*, the bacterium responsible for common infections such as scarlet fever, cellulitis, puerperal fever and post-operative wound infections. Domagk was awarded the Nobel Prize for his work in 1939, but as Hitler prohibited any German national from accepting the prize, he had to wait until the Second World War was over before collecting his award.

The next landmark discovery was penicillin, and, with a much broader range of antibacterial activity than sulphonamides, the drug was released in time to save the lives of thousands of soldiers wounded in the Second World War. Its discovery heralded the antibiotic era; now just a few tablets could cure killers like pneumonia, meningitis and diphtheria, and under antibiotic cover surgical techniques became more radical and invasive as well as safer.

The history of penicillin, from its discovery and purification through to full-scale production is one of triumph but also of conflict between the major players. The Scottish doctor Sir Alexander Fleming was the first to notice the antibacterial properties of the mould *Penicillium notatum*, although the way he made the discovery was somewhat serendipitous. During the First World War Fleming worked in battlefield hospitals on the Western Front, where every day he could only look on helplessly while thousands of young soldiers died from wound infections. This experience influenced his choice of research, and by the time of his landmark discovery he had already made his name by isolating the weak antibacterial enzyme, lysozyme, part of the body's natural defences against invading bacteria.

By all accounts Fleming was the archetypal professor—clever but forgetful and highly disorganized. His research laboratory at St Mary's Hospital, London, was always a mess, but on one particular occasion in 1928 we can forgive him. He returned from holiday to find some forgotten bacterial culture plates overrun with mould, but as he was discarding them he noticed a clear zone round some fungal colonies where the bacterial growth was inhibited. Interested, he identified the mould as a new member of the *penicillium* family and set about extracting the antibacterial substance it produced. He called this 'penicillin' and published his findings the following year.[16] But the publication attracted little interest, and as penicillin was difficult to purify Fleming had very small quantities to work with. However in the same year he tried treating a few patients with infections but had little success until one of his laboratory assistants, Dr K. B. Rogers, developed a pneumococcal eye infection just days before he was due to compete in a rifle-shooting competition, and Fleming was persuaded to try again. With penicillin the infection cleared up rapidly and Rogers was clear-sighted for the match,[17] but for some reason Fleming did not pursue this potentially important result.

In 1932 Dr Cecil G. Paine, a pathologist from Sheffield Royal Infirmary who had trained at St Mary's Hospital, also succeeded in treating eye infections by irrigating the eye with the 'juice' from Fleming's mould. Among his successes were two newborn babies with gonococcal conjunctivitis and a colliery manager with a penetrating injury of the right eye. He developed a pneumococcal infection in the eye so the desperately needed operation to remove the foreign body could not be carried out. Paine succeeded in curing the infection with penicillin and the eye was saved. But still Fleming held back; he needed a professional chemist to purify enough penicillin for full-scale clinical trials, and when he could not find anyone willing to undertake the project he lost interest, preferring to work on lysozyme.

Then, in 1938, enter Howard Florey, a bright young Australian-born doctor who headed a large research team at the Sir William Dunn School of Pathology in Oxford. He and his co-worker, the equally bright biochemist Ernst Chain, a Jewish refugee from Hitler's Germany, decided to have a go at purifying penicillin. They had succeeded in producing enough of the drug for animal testing by 1940, and against the backdrop of war in Europe and the ever present threat of Hitler's invasion of Britain, they performed experiments in mice that showed just how powerful the drug could be. Now they believed they had a vitally important drug in the making, and while the flotilla of assorted craft rescued the remains of the shattered British army from the beaches of Dunkirk, the team met to plan for every eventuality. Florey, Chain and others secreted *P. notatum* spores in the lining of their coats so that in the event of a German invasion and the destruction of the Oxford laboratory they could start up again elsewhere.[18] Fortunately invasion was averted and the work continued.

Fleming read of the results of Florey's animal experiments in *The Lancet* in August 1940[19] and promptly paid the Oxford team a visit, as he said 'to see what you've been doing with my old penicillin'.[20] This was the first contact between penicillin's discoverer and the Oxford team, and, famously, Chain is supposed to have expressed surprise that Fleming was still alive.

Florey was now seeking commercial backing for large-scale production of penicillin for clinical trials, but the climate in war-torn Europe was not conducive and in the end he set about the task himself, virtually turning the Dunn School into a factory. Clinical trials were under way by the beginning of 1941 and the results were dramatic; the condition of all six patients treated improved with penicillin and two were saved from almost certain death. With these encouraging results and with America's imminent entry into the war, Florey obtained commercial backing from the US, and mass production quickly followed. Relations between Fleming, Florey and Chain were not always the most friendly, and Florey was particularly unhappy when the press awarded Fleming the lion's share of the credit for the 'miracle cure'. However, with very different characters and skills they had each uniquely contributed to the final 'miracle' and in 1945 they were jointly awarded a Nobel Prize for their work.

With this incredible achievement all seemed set to conquer bacteria; hundreds of new antibiotics were discovered so that the whole spectrum of bacterial infections could now be cured. Some confidently predicted the demise of infectious disease microbes that had been the scourge of mankind since the advent of agriculture some 10,000 years ago—but it was not to be. With the indiscriminate over-use of antibiotics, microbes have fought back. Antibiotic-resistant bacteria like methicillin-resistant *Staphlococcus aureus* (MRSA) and *M. tuberculosis* have emerged and we are running out of options. As we shall see in the next chapter, people are again dying of infections that only a decade ago were easily treatable.

8

THE FIGHT BACK

Throughout history microbes undoubtedly had the upper hand, but by the mid-twentieth century our fight back was well under way and for a time it really seemed as if killer epidemics could be conquered at last. But no sooner had the much quoted United States surgeon General William Stewart declared that 'we can now close the book on infectious diseases' in 1967, than we began to see new and sometimes lethal microbes emerging. Since then they have hit us at the rate of around one a year, and now the frequency is increasing, a scenario that seems to mirror events of 10,000 years ago when animal domestication prompted a spate of new human infections. And the reasons today are broadly the same as they were then—environmental changes that bring us into contact with 'new' microbes which are then spread by travellers.

The major emerging microbe threats today are the unrelenting global spread of HIV, the looming spectre of microbial drug resistance and the frightening possibility of a mutated H5N1 bird flu causing a human pandemic. But in the post-genome era we are no longer battling in ignorance against an unknown enemy. In this chapter we look at the reasons behind the increasing numbers of emerging microbes, and ask whether, with all our accumulated scientific knowledge, we are better prepared to fight them than our ancestors were.

Despite all the ups and downs of our battle with microbes, the fact remains that *Homo sapiens* is the most successful species that ever roamed the planet. With our complex brains, capable of detailed forward planning and highly sophisticated speech communication, we have succeeded in adapting to life in almost every environmental niche on the Earth's surface. And despite the ravages of natural disasters, wars and killer microbes, our population has grown steadily so that we now dominate life on Earth. The world's population has roughly doubled every 500 years since the beginning of the Christian era when it stood at some 300 million. It reached 1 billion by 1800 and 1.6 billion by 1900, but the most dramatic increase occurred very recently: during the twentieth century average life expectancy doubled and the population rose fourfold.[1] Now there are well over 6 billion people in the world, and 50 per cent of us live in cities (Figure 8.1).

The result of this unprecedented population explosion is already apparent; we live in a world where our fast-disappearing

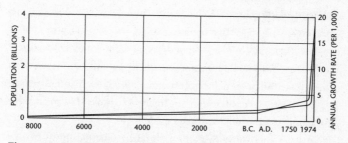

Figure 8.1 Population growth, 8000 BC–AD 1974
Source: Adapted from *History of Human Population* Ansley J. Coale. Copyright © September 1974 by Scientific American, Inc. All Rights Reserved.

natural jungles are being replaced by rapidly expanding concrete ones. The world's largest city is Tokyo with an incredible 34 million inhabitants, but now most of our huge metropolises are in the developing world, topped by Mexico City, home to over 20 million. And since our population is predicted to reach 8–9 billion by 2050 and 9–10 billion by the end of the twenty-first century, this growth of towns and cities is likely to continue unabated.

The amazing human success story comes at a high price; whereas all other species are controlled by their environment, we now control ours. So what sort of a job are we making of it? In a world of a finite size with limited resources these stark population statistics are alarming; the situation we have created is not sustainable for the future. Our burgeoning population, combined with human greed, underlies most of our current global problems: the energy crisis; the lack of clean water; air, sea and land pollution; plant and animal extinctions with the loss of biodiversity; the hole in the ozone layer and global warming. And in addition to this catalogue of potential disasters, overpopulation is key to the rise in emerging microbes.

In an overcrowded world we are always pushing against the margins of civilization. Whether in search of food, work, shelter or merely an exciting challenge, we invade new environments, disrupting ecosystems that have remained unchanged for thousands of years. Be it rainforests destroyed, rivers dammed or wild animals trapped, each is a niche for microbes we know little about, some of which have the potential to infect and even kill us. A quick glance at a list of our recently emerging microbes shows that most were first acquired from wild animals (Table 8.1).

We have seen how SARS emerged in Guangdong province where people like to select their meat live from wet markets, and tests show that some local farmers and market traders have met the

Table 8.1 A sample of human pathogens which have emerged since 1977.

Year	Pathogen	Disease	Animal of Origin
1976	Ebola virus	Haemorrhagic fever	Not known
1977	Hantaan virus	Haemorragic fever with renal syndrome	Rodents
1977	Legionella pneumophilia	Legionnaires disease	None
1982	Borrelia burgdorferi	Lyme disease	Deer, sheep, cattle, horses, dogs and rodents
1983	HIV1	AIDS	Chimpanzee
1986	HIV2	AIDS	Sooty Mangabey
1993	Sin Nombre virus	Hantavirus pulmonary syndrome	Deer mouse
1994	Hendra virus	Viral encephalitis	Fruit bat
1997	H5N1 flu virus	Severe flu	Chicken
1999	Nipah virus	Viral encephalitis	Fruit bat
2002	SARS virus	Atypical pneumonia	Palm civet cat

SARS virus before. So the virus had clearly made the transition from animal to man on several previous occasions; who knows when it or another like it will do so again?

The ecosystem of the tropical rainforest is the most diverse in the world, and it is teaming with deadly microbes. For centuries the yellow fever virus and malaria parasite prevented us from conquering the African jungle, and there are many more microbes lurking ready to pounce on anyone who interferes with its delicate balances. One famous example is the deadly Ebola virus, renowned for its explosive epidemics in remote tropical settings.

Its natural animal host is still unknown and it is imperative that we find it and stop these outbreaks because each new one gives the virus an opportunity to evolve more efficient ways of spreading between people and infecting a larger population.

HIV1 is another virus that emerged from the rainforests of Africa, where it jumped from the chimpanzee subspecies *Pan troglodytes troglodytes* which carries it as a silent infection. These large primates live in central Africa, but they have almost been driven to extinction by our destruction of their habitat and our desire for 'bush meat'. It is easy to see how a virus in chimps' blood could transfer to a human during the bloody process of killing and butchering such an animal, and analysis of banks of stored African blood samples shows that HIV1 has transferred to humans on several occasions before the 1930s, although it did not spread significantly between humans until the 1970s.[2] With tourists now demanding the bush meat experience, and some 1–5 million metric tonnes of it consumed every year in the Congo alone, it is not surprising that scientists have recently turned up evidence of several other primate viruses that have, on occasions, jumped to humans.[3] It is probably only a matter of time before one of these succeeds in spreading between humans and causes a new epidemic.

But the bush meat trade is not the only cause for concern. International trade in live wild animals for agriculture, scientific experimentation or pets is now a multibillion dollar industry. Germany experienced the first outbreak of Marburg fever, caused by an Ebola-like haemorrhagic fever virus, which arrived in 1967 in a consignment of African green monkeys from Uganda and infected thirty-one laboratory workers, killing seven of them. More recently a monkeypox outbreak in the US began when the virus was imported from Ghana along with Gambian giant rats for the exotic pet market.[4] The virus transferred to prairie dogs housed in the same pet shop as the rats and jumped from there to their owners. The chain of infection was only terminated after the microbe had infected seventy-one people. And there are reports that farmed crocodiles in Papua New Guinea fed on wild pig meat have developed trichinella, a life-threatening worm that they could easily pass on to their keepers.[5] Who knows what else might be hiding just round the corner?

Poverty

It is glaringly obvious from a glance at the figures that poverty is *the* major cause of microbe-related deaths. On a worldwide scale microbes are still major killers, accounting for one in three of all deaths. But the huge discrepancy in the death rates between rich and poor nations reveals the stark reality. Whereas only 1–2 per cent of all deaths in the West are caused by microbes, this figure rises to over 50 per cent in the poorest nations of the world, and it is in these highly microbe-infected areas where over 95 per cent of the global deaths from infections occur.[6] Most of the 17 million killed by microbes each year are children in developing countries where the link with poverty is clear. It is the poor who are malnourished, live in filthy, overcrowded urban slums and go without clean drinking water or sewage disposal, and therefore they are the ones who fall prey to the killer microbes: HIV, malaria, TB, respiratory infections and diarhoeal diseases like cholera, typhoid and rotavirus; all eminently preventable and treatable given the resources.

The spread of HIV is an excellent example of how microbes exploit the poor, striking at the most disadvantaged in the community. The virus emerged in Central Africa and spread silently throughout the continent in the 1970s, given a head start by its long silent incubation period, and aided by despotic leaders, corrupt governments, civil wars, tribal conflicts, droughts and famines. Carried by undisciplined armies and terrorists, the virus infiltrated city slums, infected commercial sex workers, was picked up by migrant workers and passed on to their wives and families. While malnutrition accelerated the onset of AIDs, breakdown of health-care services in the political turmoil of Africa excluded any possibility of medical support for the millions in need.

Now we are living through the worst pandemic the world has ever known, with 40 million living with HIV, 25 million already dead and around 10,000 dying daily—the equivalent of over three 9/11 disasters every twenty-four hours. A third of people living in sub-Saharan African cities are HIV-infected, and while highly active antiretroviral therapy (HAART) has converted this lethal disease into a manageable chronic infection in the West, presently only a tiny proportion of Africans living with HIV receive this treatment; for most there is no hope of obtaining the drugs vital for keeping them alive.

The dynamics of HIV in Africa reflects its mode of spread. As the virus is sexually transmitted gender inequalities mean that women are particularly vulnerable. In general they are poorer and less well educated than their male counterparts, and are often powerless to choose or restrict their sexual partners, or to insist on condom use. Indeed many are forced to exchange sex for essentials like food, shelter and schooling. Now one in four African women are HIV-infected by the age of twenty-two years (compared to one in fourteen men of the same age), and women account for 60 per cent of all those living with HIV.

Over 90 per cent of HIV-positive women in Africa are mothers, and the virus has created 15 million orphans worldwide, 12 million of them in sub-Saharan Africa. These children are bearing the burden of the HIV pandemic; they miss school to care for their sick mothers or to earn the family income; the virus has not only deprived them of their parents but their childhood and their education as well.

Travel

All through history we have seen how travellers spread microbes: first causing the merger of discrete infectious disease pools in the Old World, then spearheading their invasion of the New World, and finally disseminating microbes to all the world's most isolated communities. Generally speaking epidemics can only move as fast as humans can travel, so in the days when commercial goods were carried on horse- or camelback along ancient trade routes, and wind-powered galleys sailed from port to port, progress was slow. Many potential human microbes must have died out along the way through lack of susceptible victims to sustain their chain of infection. But when the transport revolution of the nineteenth and twentieth centuries speeded things up microbes also benefited, spreading further and faster than ever before. This is nicely illustrated by the shrinking of travel time between the UK and Australia, which telescoped from a whole year by sailing ship in the eighteenth century, to 100 days by clipper in the early nineteenth century, to fifty days by steamer at the beginning of the twentieth century.[7] This significantly increased the chance of travellers spawning an epidemic like measles, which, with an incubation period of fourteen days, had to pass through a chain of six susceptible people on board a clipper to reach Australia in the nineteenth century but only three when travelling by steamer in the early twentieth. The contraction of travel time was felt particularly in small islands like Fiji and Iceland where the sparse population relied on shipping links for vital supplies. Too small to sustain endemic microbes themselves, epidemics arrived by sea, and their frequency increased as travelling time decreased.

But the effect of advances in shipping on microbe spread was negligible compared to the impact air travel had in the twentieth century. This collapsed geographical space so drastically that now we can get from virtually any large city in the world to any other in a day, and with our huge global population, bigger aircraft, and low-cost airfares, more people are travelling more frequently and further than ever before. An interesting personal study by David Bradley of the London School of Hygiene and Tropical Medicine in the late 1980s conveniently makes the point. He plotted the lifetime travel patterns of four male generations in his family (his great-grandfather, grandfather, father and himself) and showed that the spatial range increased tenfold with each generation, his own being 1,000 times that of his great-grandfather.[8]

Microbes were not slow to exploit this new opportunity. With millions of people taking to the air every year, the risk of transporting microbes, be they inside a human, animal or insect, increased dramatically. We have seen many examples in previous chapters: West Nile fever virus probably jetted into the US from the Middle East inside a stowaway mosquito, HIV arrived in the US inside tourists from Haiti and then city-hopped throughout the US and Europe, and SARS made the leap from Hong Kong to five other countries inside human incubators. Even traditionally tropical infections are now jet-setting around the world. In 1983 the landlord of a village pub 12 miles from Gatwick airport near London suddenly collapsed with malaria, and the microbe also infected an unfortunate motorcyclist who happened to pass through the village at the time.[9] And more recently several people living near Geneva airport caught malaria without ever leaving Switzerland when parasite-laden mosquitoes hitched a ride in a plane from the tropics.

Antibiotic Resistance

In 1945 Sir Alexander Fleming predicted that microbes would evolve antibiotic resistance, pointing out in his Nobel Prize speech that 'It is not difficult to make microbes resistant to penicillin in the laboratory by exposing them to concentrations not sufficient to kill them, and the same thing has occasionally happened in the body.'[10] At the time hardly anyone heeded his warning, perhaps because as soon as penicillin hit the market scientists set to work uncovering more natural antibiotics while chemists tinkered with the natural molecules to make new active derivatives. So with literally hundreds of wonder drugs to choose from few doctors thought very seriously about microbe resistance. Sixty years later, we have a huge global problem: increasing numbers of microbes are acquiring resistance to multiple drugs, the natural source of antibiotics has virtually run dry, and there are few new drugs in the pipeline. How have we reached this crisis situation?

Most of our antimicrobial drugs, such as penicillin, are antibiotics—natural chemicals produced by bacteria and fungi to ward off other microbes. The word antibiotic, meaning 'against life', was coined to describe their power to kill microbes by shutting down essential functions. The beauty of this from the human point of view is that antibiotics only target microbes, leaving our own cells unharmed. Penicillin works by blocking the transpeptidase enzymes that are essential for building a bacterium's tough outer wall. So when bacteria try to grow in the presence of penicillin their cells rupture and die. But with their rapid reproduction rate bacteria are expert at adapting by natural selection. A chance genetic change that confers resistance to a drug will give an individual microbe such a competitive edge that within hours its offspring, all similarly resistant, will outstrip the competitors and dominate the population. But many bacteria don't have to wait to inherit a resistance gene; they can pick it up from others by gene swapping, a process that is rife among bacterial communities. Since many antibiotic-resistant genes are extra-chromosomal, carried either by plasmids or other transposable elements, they are highly mobile. So multidrug resistance is not only spread by the rapid growth of resistant bacterial clones but also by swapping of the offending genes.

In recent years our use of antibiotics has escalated, some legitimate and some irresponsible and inappropriate, and microbes have been quietly cashing in on the situation. New and complex surgical procedures often require antibiotic cover to give the patients the best chance of survival, while at the other end of the scale antibiotics are regularly prescribed for minor ailments like throat infections that are usually caused by viruses and so are unlikely to benefit. Then, most of us are happy to take medicines while we are ill, but once we feel better it is tempting not to complete the course. This partial treatment is bound to generate antibiotic resistance when the drug kills the most sensitive microbes first, leaving the resistant ones to grow unimpeded. The problem is compounded by over-the-counter antibiotics, with people free to choose the type of antibiotic and length of treatment. No wonder the highest levels of multidrug-resistant microbes are found in countries with easiest access to the drugs. The community microbe, *Streptococcus pneumoniae* (generally known as pneumococcus), is a case in point. It heads the list of bacterial infections and deaths worldwide, causing bronchitis, ear and sinus infections as well as life-threatening pneumonia, meningitis and septicaemia. For years pneumococcus was successfully killed by penicillin, but emerging penicillin resistance is now a worldwide problem. While the incidence of resistance has risen from 5 per cent to 35 per cent in twenty years in the US, where antibiotics are freely available, it has remained at around 5 per cent in the UK.[11] But scientists have fought back with a new pneumococcal vaccine which was used for the first time in the US in 2000. This reduced childhood infections, and, interestingly, the consequent reduction in antibiotic use reversed the rising tide of resistant microbes by removing their selective advantage.

Drug resistance is also potentiated by the enormous overuse of antibiotics in agriculture. Farmed animals consume over half the world's antibiotics. These are used not just to treat a sick animal, but the whole herd gets a dose to prevent the microbe spreading. Worse still, by some mysterious means low-dose antibiotics promote the growth of farmed animals, so they may be added to animal feed. And although some countries have now banned this practice, in others herds still receive lifelong treatment. This is a recipe for emerging resistant microbes, and indeed a multidrug-resistant strain of the zoonotic microbe *Salmonella enterica typhimurium* that causes diarrhoeal disease in millions of people every year has already emerged in animals and spread to humans.

Staphylococcus aureus lives harmlessly in the noses of many healthy people, but causes havoc when let loose in hospital surgical wards and intensive care units. It can live for months in bedclothes or dust and is often inadvertently carried from one patient to another by hospital staff. It targets the most debilitated patients, infecting their lungs, surgical wounds and catheter sites, and from there it can invade the blood to cause a highly dangerous septicaemia. The microbe's first manoeuvre in the fight against antibiotics was to produce lactamase, an enzyme that destroys the penicillin molecule. When this penicillin-resistant *Staph. aureus* became a problem doctors switched to using one of its semi-synthetic derivatives, methicillin. But the bacteria then produced a new

type of transpeptidase that could not be blocked by either penicillin or methicillin. So to treat this methicillin-resistant *Staph.
aureus* (MRSA) doctors turned to vancomycin, a drug that dis-
rupts the bacterial cell wall by a different route. Known as 'the
antibiotic of last resort', this worked for thirty years, but in 2002
the microbe struck back and vancomycin-resistant MRSA
emerged in the US; its global spread is now a real possibility.

And so while the arms race continues MRSA plagues our
hospitals, closing wards and cancelling operating lists. In the
UK, MRSA levels rose dramatically in the mid to late 1990s
(Figure 8.2), with the additional cost to the health service of £1
billion annually. But contrary to popular belief doctors still have
one or two antibiotics up their sleeves to treat MRSA, and there
are a few new drugs coming on line. Also MRSA can be controlled in other ways. Hospitals in Denmark, Sweden and Holland

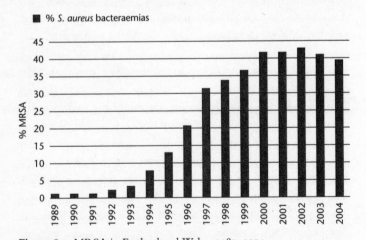

Figure 8.2 MRSA in England and Wales, 1989–2004

Source: D. Livermore, 'Is Antibiotic Resistance an Unwinnable War?', Wellcome Focus,
pp. 16–17, 2005, by permission of the Wellcome Trust.

have conquered the microbe by scrupulous hand-washing and strict patient isolation, but this requires more expensive isolation facilities than many countries can afford.

MRSA is by no means the only 'superbug'. On a worldwide scale multidrug-resistant TB and malaria are much more serious problems, and now drug-resistant HIV is raising its ugly head. Although caused by completely unrelated microbes using quite different transmission routes, these three persistent microbes are inextricably linked. All three target sub-Saharan Africa, where HIV potentiates the effect of the other two, particularly reactivating quiescent TB. The scale of the problem posed by each microbe is enormous and the only answer is cheap and effective vaccination. But with no vaccines yet available for HIV or malaria, and the traditional BCG vaccine for TB only partially effective, control programmes rely heavily on drug treatments which are now jeopardized by microbe resistance.

Antibiotics are not effective against viruses so the emergence of HIV in the 1980s prompted pharmaceutical companies to develop antiviral drugs, and with new molecular know-how for identifying drug targets the field blossomed. In 1990 there were only four on the market; fifteen years later we have over forty, most of which are aimed at HIV. These drugs revolutionized the management of HIV infection in the West and now, with cheaper drugs available from the pharmaceutical companies, and generic manufacture, HAART is being slowly rolled out in developing nations. But still we cannot eliminate the virus or cure the sufferers. These drugs simply buy time until eventually patients run out of treatment options and then the microbe can grow unimpeded.

There are all sorts of reasons why people living with HIV do not always take their drugs regularly. For one thing it is a lifelong commitment and often involves many different tablets taken several times a day. Also the drugs may produce unpleasant side-effects and when these are compounded with the symptoms of HIV infection, particularly in its late stages, the drugs are often not tolerated. With this non-compliance it is no wonder that HIV, like other microbes, is evolving drug resistance. Doctors used to think that resistant strains were weaker and less likely to be transmitted than their non-mutated counterparts, but a recent report describing a promiscuous gay man in New York who was infected with a highly virulent HIV strain suggests otherwise.[12] Labelled 'AIDS Superbug' by the press, the virus is resistant to three of the four classes of anti-HIV drugs and is almost impossible to treat effectively. Worse still, his disease progressed exceedingly rapidly so that twenty months into the infection the man's immunity was as low as that typically seen ten years from the start. It remains to be seen whether this superbug is on the increase, but this case stresses the importance of prevention strategies as well as treatment.

Clearly vaccine development is top priority in the battle against HIV, but although scientists have tested products that stimulate either antibody or killer T-cell responses, neither can protect against the virus. Since HIV can outstrip natural immunity by mutating rapidly, maybe a conventional vaccine will not succeed. Other strategies, like producing a therapeutic vaccine to boost immunity and enable virus carriers to remain symptom free, are being tested, but no one is optimistic that a vaccine will be available in the next five years.

In the meantime we must fight the virus without a vaccine by attacking it on all fronts. If we can interrupt its chain of infection and reduce R_0 to less than one then we will eventually win, but to have any chance of this we need to ramp up all the public health measures: education programmes targeted at high-risk groups, needle exchange facilities for IV drug users, and free condom distribution. In Africa

empowerment of women to take control of their lives is an imperative if the virus is to be halted, and to this end microbicidal vaginal
creams are being developed but to date none has proved protective.

M. tuberculosis evolved resistance to individual anti–TB drugs soon
after drug treatment began in the 1950s, but it was so successfully
controlled by multidrug treatment regimens that some confidently predicted global elimination of TB. But this was not to
be: M. tuberculosis began to get the upper hand in the 1970s and by
the early 1990s it was clear that we had a global emergency.

The world was first alerted to this rapidly growing pandemic by
a completely unexpected outbreak of TB in New York. In 1968
the Mayor of New York was planning to clear TB from the city,
but just ten years later it was on the brink of an epidemic. Starting
from familiar territory in the deprived inner-city districts of Central Haarlem and Lower East Side, the microbe crept out to
surrounding areas, eventually colonizing all the New York health
districts with the sole exception of the very wealthiest. The
epidemic peaked in 1992 when the worst affected areas reported
well over 100 new cases per square mile.[13]

It soon became apparent that this alarming resurgence of TB was a
worldwide phenomenon caused by a combination of increasing
poverty and homelessness (accompanying the cutbacks in the public
health expenditure in US cities in the 1980s), together with HIV-
induced immunosuppression reactivating quiescent TB, and drug
resistance. Now despite BCG being the most widely used vaccine in
the world, an estimated 2 billion people are infected with
M. tuberculosis—one third of the world's population. Thankfully
most of these infections are inactive, but worldwide, 8 million people
have active TB and 2 million die of it every year (Figure 8.3). Not
surprisingly, over 95 per cent of the problem is in the developing

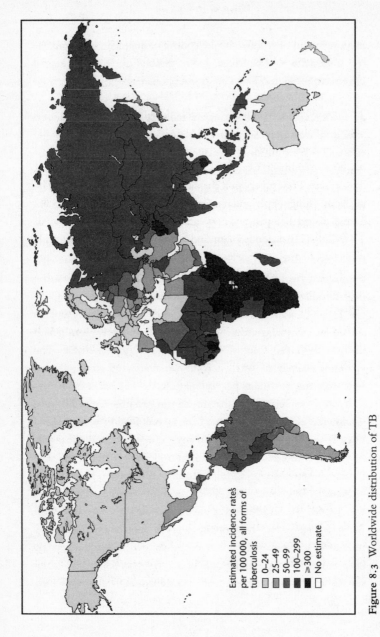

Figure 8.3 Worldwide distribution of TB

Source: Reprinted from *The Lancet 367*, Christopher Dye 'Global Epidemiology of Tuberculosis', p. 398, © 2006, with permission from Elsevier.

Estimated incidence rates per 100 000, all forms of tuberculosis

0–24
25–49
50–99
100–299
≥300
No estimate

world, with sub-Saharan Africa being the worst affected. Here at least 11 million people have the lethal combination of TB and HIV, and since TB generally affects adults in their most productive years of life, the microbe has an enormous economic impact.

Unlike MRSA, *M. tuberculosis* cannot pick up resistance genes from other microbes so it relies on spontaneous mutations to outwit the drugs. And since mutations are rare, the double mutation it needs to develop multidrug resistance can realistically only have arisen from drug misuse. Poor supervision and compliance, inconsistent prescribing, erratic drug supply, and unregulated, over-the-counter drug sales have all played their part, and now mutant multidrug resistant strains of TB (MDR-TB) account for around 10 per cent of new cases worldwide. Worst hit are Russia and the former Soviet Union states due to the breakdown of health-care services during the political upheavals of the 1990s, and sub-Saharan Africa with its enormous HIV problem. The only solution we have is longer treatment regimens with more toxic drugs, but at 200 times the price of the conventional treatment they are just not available where they are most needed.

The fight against the TB pandemic began with the WHO directly observed therapy—short course (DOTS) programme in the 1990s. By ensuring patients were seen by medical staff and had a regular supply of free drugs, this aimed to give sufferers proper treatment, allowing the microbe no chance to develop resistance. The programme worked well for drug-sensitive TB but most cases of MDR-TB do not respond. Now the problem in South Africa seems to be spiralling out of control. Poor treatment of MDR-TB has already led to the emergence of even more drug-resistant XDR-TB (extensively drug-resistant TB), and if nothing is done to stop this escalation, we will soon have an epidemic of completely drug-resistant TB on our hands. And with new drugs and more effective vaccines just a distant dream, this will be disastrous.

The early twentieth century saw the malaria parasite banished from the US and most of Europe, and in the 1950s and 1960s WHO launched a global eradication programme. Using the insecticide DDT (dichloro-diphenyl-trichloroethane) to kill off the mosquito vector, and chloroquine to treat sufferers, the scheme made some significant inroads, particularly in South America, India, Sri Lanka and the former Soviet Union, although less effort was made to control the microbe that was rampant in sub-Saharan Africa. But the global programme soon ran into problems: it was vastly expensive, people resented the repeated insecticide-spraying of their homes, and when DDT-resistant mosquitoes emerged it was the final straw. For several decades, while other killers were being tackled successfully, the malaria parasite was allowed to run riot. The world malaria burden remained unchanged until the beginning of the twenty-first century when deaths from malaria began to rise. The parasite now infects around 500 million people in 103 countries every year and kills 1 million—one death every thirty seconds. The casualties are mostly children in sub-Saharan Africa where 90 per cent of the world malaria burden falls (Figure 8.4); in the worst affected countries one in every four children die before the age of five years.

In this recent resurgence the malaria parasite has benefited from wars, civil unrest and weakened health-care services. Changes in the environment and climate have brought malaria-carrying mosquitoes to new areas, while massive population growth has facilitated parasite spread. Insecticide resistance also remains a problem, but the major factor in this evolving disaster is the emergence of drug-resistant parasites. Chloroquine was once the mainstay of malaria treatment in sub-Saharan Africa, but now resistant microbes are widespread and doctors working in the killing fields have nothing else—they are powerless to treat the disease.

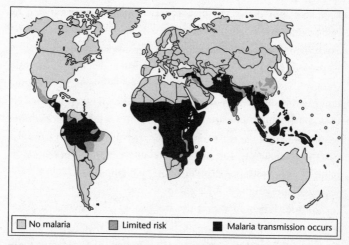

No malaria | Limited risk | Malaria transmission occurs

Figure 8.4 Geographical distribution of malaria worldwide

Source: Reprinted by permission from Macmillan Publishers Ltd: *Nature Outlook*, p. 926 ©2004.

Despite the gloomy figures and forecasts there are reasons for optimism in our battle against malaria. To have any effect we need to break the microbe's transmission cycle and there are several new tools in the pipeline which will help. Satellite imagery can map malaria hot spots in Africa, a mutant form of *A. gambiae* which cannot transmit malaria is being prepared for release and potential vaccines are in clinical trials. But scientific advances are only part of the answer to the conquest of malaria in Africa. Insecticide-impregnated bed nets, cheap and effective against malaria, are used by less than 2 per cent of African children, and new anti-malaria drugs extracted from the herb qinghao (*Artemisia annua* or sweet wormwood, an ancient Chinese anti-fever herbal remedy) are available in South-east Asia to treat chloroquine-resistant disease, but not in most of Africa.[14] What is needed to fight this most persistent of parasites is money: an estimated \$5 billion dollars annually for the foreseeable future, not only to fund essential prevention and treatment programmes in Africa but also to provide local people with the education and the infrastructure to help themselves.[15] A tall order? Yes, but if neglected then the combination of drug resistance and spread to new areas aided by global warming will cause malaria cases to rocket, millions more lives will be lost, all the exciting new scientific advances from the malaria genome project will go to waste, Africa will not attain its economic potential and the microbe will triumph.

Flu

Ever since the Chinese domesticated water fowl and pigs some 9,500 years ago new strains of flu have probably jumped to humans causing epidemics and more recently pandemics. Unlike most other microbes that jumped species all those years ago, flu continues to jump at regular intervals, and to date modern science has had little impact on its exploits.

Human-adapted flu viruses can circulate continuously among us, generally keeping a low profile, but causing outbreaks each winter that kill around a quarter to half a million people globally, mostly targeting the elderly and chronically sick. But every now and then a pandemic strain appears that sweeps round the world infecting and killing millions. Of the three pandemics in the twentieth century, by far the worst was the famous 1918 Spanish flu which hit as the First World War was ending. Just as in ancient times, the pandemic enveloped an unsuspecting world like a great tidal wave, infecting half the world's population and leaving 20–50 million dead in its wake. People had no idea where it came from, what caused it or how to combat it. Now, less than 100 years later, we know the answers and are in a position for the first time to watch a possible new pandemic unfold.

After an attack of flu we are protected from further infection, mostly by antibodies directed against H (haemagglutinin) and N (neurominidase) proteins on the surface of the virus particle. So after an epidemic, when most people are immune, the virus loses its power. But unlike other viruses that cause acute infections, the flu virus fights back by changing its H and N genes so that it can dodge our immune system and re-infect. There are fifteen different H genes and nine N genes, and flu virus strains are named by their particular combination. The original Spanish flu was caused by H1N1 virus, which was replaced by H2N2 in 1957 (Asian flu), and in 1968 H3N2 (Hong Kong flu) emerged, then in 1977 H1N1 (Spanish flu) reappeared.

Domestic and wild birds are the natural reservoir for flu viruses, carrying a whole variety of different strains in their guts as a silent infection, and excreting them in their droppings. And because the virus has a segmented genome with eight separate genes, those from different strains sometimes get mixed and matched to create 'new' strains. This happens when two different strains of flu virus infect the same cell and hybrid viruses emerge containing an assorted mixture of genes from the two parent viruses.

Bird flu viruses usually lack the receptor binding protein needed to infect human cells, but some domestic animals like pigs and horses are susceptible to both bird and human strains. So gene swapping between human and bird strains often occurs in pigs or horses, causing a major genetic change in the virus make-up called an *antigenic shift* (Figure 8.5). Occasionally after this mixing a 'new' virus strain emerges that can infect and spread in humans, and as the population is completely naïve to this 'new' strain it can spark a pandemic.

After a flu pandemic the same virus strain generally remains in the community for a while, and as it circulates it slowly

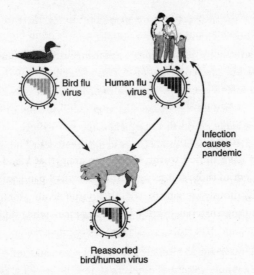

Figure 8.5 The emergence of pandemic flu virus strains after reassortment in a pig

Source: Dorothy Crawford, *The Invisible Enemy* (2000), by permission of Oxford University Press.

accumulates mutations. This is called *antigenic drift,* and when the virus has changed sufficiently to be unrecognizable to our immune systems then it will be able to infect again and cause a new epidemic.

H5N1 bird flu is not new. It was first recognized in chickens in Scotland in 1959, and turned up next on South-East Asian chicken farms in the 1990s. At this stage it only caused a mild disease in poultry, but some time in the mid-1990s it mutated into a highly virulent strain that killed virtually all infected chickens in forty-eight hours. This strain first jumped to humans in Hong Kong in 1997 when it infected eighteen people, killing six of them. In an attempt to prevent its spread the authorities ordered the slaughter of millions of chickens, which seemed to do the trick since all went quiet for a while, but in 2003 the virus resurfaced in Hong Kong, Vietnam and Thailand. Now we have a wholesale bird flu pandemic affecting poultry flocks in at least eighteen countries. So far it has only attacked humans who handle infected birds and is not yet adapted for transmission between humans, but the alarming fact is that it has killed over half the people it has infected.

To try to figure out why H5N1 is such an efficient killer scientists have turned to the H1N1 flu strain that killed around 2.5 per cent of those it infected during the 1918 pandemic. They have ingeniously reconstructed this virus from stored post-mortem lung material from a US serviceman who died in the pandemic, and also from an Alaskan flu victim whose body was preserved by being buried in the permafrost for nearly 100 years. In the classical scenario outlined above bird flu strains swap genes with human strains before they can infect humans efficiently, and this genetic mixing often takes place in domestic pigs. Scientists working on the lethal 1918 H1N1 strain still don't know where it came from, and some even speculate that it jumped straight from birds to humans, bypassing any intermediate host.[16] They then found that compared to a non-pandemic flu strain it had undergone ten mutations that allowed it to infect and grow very efficiently in human cells. They pinpointed one particular mutation in its NS1 gene that prevents infected cells from producing a cytokine called interferon, one of the body's first lines of defence against viruses. So immediately the virus is one step ahead, multiplying rapidly in the lungs and producing up to an amazing 39,000 times more virus than its non-mutated relatives.[17] The body reacts to this rampant infection with a massive and inappropriate inflammatory response called a 'cytokine storm', and the unfortunate victim literally drowns as the air sacs of their lungs fill with blood and fluid. H5N1 has already acquired this particular NSI mutation[18] and several of the others found in pandemic H1N1, which is why it is so lethal. All it needs now is to acquire mutations that help it infect and transfer between humans more easily, so it is probably only a matter of time before it causes a human pandemic.[19]

It is difficult to prepare for a possible pandemic some time in the future without knowing the exact genetic make-up of the culprit virus. Although we have antiviral drugs that can combat the flu strains presently circulating, it is by no means certain that they will help to control a new strain. Similarly vaccines made to prevent today's H5N1 may not protect against a mutated variant. So we are on a knife edge between on the one hand waiting to discover the exact molecular make-up of the future pandemic strain and being too late to respond, and on the other making early preparations that may in the event prove unnecessary or ineffective. At this stage it is imperative to monitor this rapidly evolving pathogen in both animals and humans, and WHO have a network of laboratories worldwide continually monitoring human flu strains to produce an appropriate vaccine each year. In collaboration with the World Organisation for Animal Health and the United Nations, staff in these laboratories are now monitoring the evolution and spread of H5N1 strains in both birds and humans.

Despite uncertainty surrounding a human H5N1 flu pandemic, there is no doubt that we are in the midst of the worst ever recorded flu pandemic in birds. The virus started life as a harmless infection in the intestines of wild birds and jumped to domestic chickens in the 1990s, where modern intensive farming techniques gave it the opportunity to adapt and evolve. When a mutated virus emerged that could infect all the organs of a chicken it killed rapidly. And now this virulent strain has not only crossed back into wild fowl but has increased its host range to include other birds (crows, pigeons, falcons, buzzards) and even some mammals such as cats. Thousands of wild birds are dying of the infection and there are real fears for species already endangered by loss of habitat, particularly if they have a limited geographical range.

Qinghai Lake in North Central China lies at the crossroads of several bird migration pathways and every year bar-headed geese are among those that drop in for a well-deserved rest after their flight over the Himalayas from India. For 6,000 of these geese that arrived in 2005 it was their last journey. They died of H5N1 beside the lake—a loss of a tenth of the world's population.[20]

If a virus that kills so rapidly relied on migrating birds for its dissemination it would not get far and the pandemic would die out. But now there is evidence that this virus, so lethal to chickens and other birds, has evolved to be harmless in ducks. Once infected they remain healthy and just shed the virus in their droppings for a week or two, quite long enough to carry it on their migration and pass it on to other susceptible birds. In theory silently infected ducks could spread the virulent virus round and round the globe, but in practice we really don't know what the future holds; with a virus so prone to mutation anything could happen. So what can we do to stop it causing further damage to our beleaguered wildlife? It is not possible to kill all infected wild birds or vaccinate susceptible ones, but we can try to prevent infection in poultry flocks with vaccination. This would at least limit opportunities for the virus to evolve, hopefully save the livelihood of many poor smallholders and perhaps prevent further passage to humans.

We have seen in this chapter how modern technology has intervened to improve the lot of the afflicted as microbes emerge, re-emerge and evolve in the twenty-first century. But we are still left with many killer microbes on the loose for which we have not found solutions. Microbes have been quick to exploit our global society, but unfortunately we have not yet come up with global solutions to control them.

CONCLUSION: LIVING TOGETHER

Microbes were the first life forms to evolve on planet Earth and now there are more of them than any other organisms, inhabiting every conceivable niche, including the bodies of other species. We relative newcomers to the planet emerge from the safe environment of our mother's womb pristine, untouched by infectious microbes, but within hours our bodies are colonized by swarms of them, all intent on living off this new food source. From then on there is no getting away from them; they surround us, living on our skin and inside our bodies in their millions. The vast majority of these microbes are either necessary for our existence or completely harmless. Only a few live as parasites, dining on our tissues and in so doing cause disease.

Throughout our history pathogenic microbes have exploited our cultural changes, turning each one to their own advantage. So while we evolved from hunter-gatherer to modern city-dweller, microbes kept us company. At each new step microbes were ready to pounce, usually jumping from their natural animal hosts and then evolving with us, generally to mutual benefit. The increasing complexity of our social structures accentuated inequalities until the divide between the rich and poor, the have and have-nots, and the vulnerable and protected, became embedded in our culture. Consequently, fuelled by overcrowded, unhygienic living conditions, and spread by travellers, opportunistic microbes caused devastating epidemics which we have only recently learnt to control. With public health measures to inhibit their spread, vaccines to deny them access to our bodies and antimicrobial drugs to kill them off, global deaths from infectious disease finally began to fall in the early 1900s.

But despite all our efforts microbes are still a major killer, accounting for 17 million deaths a year, around a third of all deaths worldwide. And as HIV took hold in the 1980s, the death rate from infections rose again, doubling between 1980 and 2000. With an increasing number of new microbes emerging, microbe drug resistance on the increase, and the threat of bioterrorism, we could be heading for a crisis.

Louis Pasteur was the first to point out that 'The microbe is nothing, the terrain everything'. We have always been part of microbes' terrain, but by taking control of Earth we have invaded their space and disrupted their natural cycles, and now we are suffering the consequences. So how can we protect ourselves from their devastating effects in the twenty-first century?

The elimination of the smallpox virus was a major triumph in our fight against killer microbes, saving millions of lives. And now measles and polio viruses are well on their way to extinction, but global elimination is not an achievable, or even desirable, goal for most pathogenic microbes. Similarly, the invention of a superdrug, the so-called 'gorillacillin' that would protect against a wide range of bacteria, is just a pipe dream.[1] And even if its production was feasible this sledgehammer approach would kill indiscriminately, wiping out the good microbes along with the bad. As free-living organisms, many bacteria form part of mutually dependent colonies which we disrupt to our peril. Those in our gut, for example, generally help to keep us healthy but may cause nasty infections if given the chance to invade our tissues. A short course of oral antibiotics is enough to disturb their microenvironment by killing off most of the susceptible ones, and this clearout often induces diarrhoea and on occasions thrush by allowing the normally harmless fungus *Candida albicans* to flourish.

Our slow rate of evolution is no match for the diversity and rapid adaptability of microbes, so we must accept that, in the short term at least, they will constantly outsmart us. Indeed having spent millions of years counteracting the very microbial products from which most of our antibiotics are derived, it is a fair bet that microbes will find ways of resisting any new products we throw at them.

The best defence we have against microbes is our brains, which can surely work out how to live in harmony with the microbes we know, and find non-disruptive ways of combating those that emerge in the future. Fortunately the genome era is upon us and our knowledge of microbes has escalated, vastly improving our understanding and ability to counteract them. Just look at the speed at which new scientific facts emerged during the SARS epidemic: the culprit coronavirus was isolated within weeks and its genome sequence mapped less than a month later. Almost immediately tests became available to diagnose suspected cases, trace contacts and identify the probable animal host, and work is now progressing fast towards a vaccine. Just a decade ago all this would have taken years to achieve. So how does the genome revolution help in our attempt to live in harmony with microbes?

The first pathogenic human virus to be completely sequenced was the Epstein–Barr virus in 1984, followed by the first bacterial sequence, *Haemophilus influenzae*, in 1995. Now several thousands of bacterial and viral sequences are known, and the comparatively huge malaria parasite genome sequence has just been completed. This provides a seemingly limitless fund of information which can surely be used to improve health globally. As we have seen in earlier chapters, the sequence of microbe genes alone can tell us about their origin and evolution as they radiated out from their source, and can uncover the molecular details of their parasitic lifestyle. But with the whole human genome sequence comprising around 25,000 genes at our fingertips, we can now look at the ways microbes interact with their human host at the molecular level. This is revealing the secrets of our genetic susceptibility and resistance to microbes, identifying drug targets, and opening up a range of exciting new vaccine possibilities.

Vaccination was the first successful form of immunotherapy, the process of harnessing the immune system to control microbes, and everyone agrees that it is still the best way forward. Indeed there is clear evidence that vaccines can reduce the rising tide of antibiotic resistance. As we saw in Chapter 8, the incidence of antibiotic-resistant pneumococcus in the US rose from 5 per cent to 35 per cent over twenty years, but when a pneumococcal vaccine was introduced in 2000, fewer antibiotics were needed, the selection pressure was removed and the incidence of resistance fell.

Traditional vaccines are made from killed or weakened microbes that induce immunity without disease. But, as we have seen with HIV, this approach does not always work, particularly if the body's immune system is suppressed by infection or disease. So now scientists are devising new and sophisticated types of immunotherapy as safe, non-invasive, 'natural' treatments for infections. Antibodies are a key part of the body's defence against invading microbes, particularly bacteria. Now designer antibodies directed against individual microbes can be made in the laboratory and used to neutralize an invading microbe. Similarly, killer T-cells, our main defence against viruses, are being harvested from healthy people and used to treat infections in people with suppressed immunity, such as transplant recipients and cancer patients. Hopefully in the future immunotherapy will be widely available to complement the more traditional ways of combating infections.

We have seen how antibiotic resistance genes and emerging microbes such as HIV and SARS jetted round the world with amazing speed, and with our present propensity for international travel other microbes will certainly follow in their footsteps. Wherever a new microbe emerges in the future it cannot be regarded as just a local problem. In deciding to hush up the SARS outbreak in Guangdong province in 2002, the Chinese government gave the virus a head start and allowed it to spread globally. And while medical personnel around the world struggled with this previously unknown disease, Chinese doctors already had successful protocols for its containment that could have saved many lives. But China is not alone in preferring secrecy. When a middle-aged couple from New Mexico turned up in New York with bubonic plague in 2002 (probably caught from a wood rat lurking in their backyard), the outside world only got to hear of it after the victims were well on their way to recovery. Perhaps it is understandable that governments wish to avoid the inevitable economic collapse that accompanies the rumour of an epidemic these days, but in our globalized world this is not acceptable. Only global cooperation can prevent the looming catastrophe of a flu pandemic. Microbes know nothing of countries and do not respect their boundaries. As Anthony Fauci, Director of the National Institute of Allergy and Infectious Diseases, USA, said when referring to our fight against HIV, 'history will judge us as a global community'. After all, that is how we have always been seen by our deadly companions.

Glossary

ague old term for malaria derived from the Latin *febris acuta*

algal bloom a rapid increase in cyanobacteria in lakes, rivers or coastal waters often caused by pollutants such as nutrients in run-off from farmers' fields

***Anopheles* mosquito** genus of mosquitoes including *A. gambiae*, the main malaria vector in Africa

anthrax a zoonotic disease caused by the spore-forming bacterium *B. anthracus*; infection may be by inoculation causing cutaneous anthrax or by inhalation causing respiratory anthrax, also called wool-sorter's disease

antibiotic a substance produced by micro-organisms that can inhibit or kill susceptible microbes

antibody blood protein produced in response to a foreign antigen that can inactivate certain infectious agents

antigenic drift alteration in the genetic code of flu virus over time by the accumulation of point mutations

antigenic shift alteration of the genetic make-up of flu virus by reassortment of genes

Archean era the early part of the Precambrian era characterized by the absence of life

avirulent without pathogenic effect

bacteriophage (phage) viruses that infect bacteria; lytic phage cause lethal infections of bacterial cells

bacterium single-celled organism with a simple, prokaryotic structure

BCG bacilli Calmette-Guerin, the attenuated vaccine strain of *Mycobacterium tuberculosis*

bejel a non-venereal syphilis-like disease endemic in Africa, western Asia and Australia

binary fission cell division to produce two identical offspring

biodiversity diversity of plant and animal life

Black Death pandemic affecting Europe, Asia and North Africa between 1346 and 1353, supposedly caused by *Yersinia pestis*

Botrytis infestans original name for *Phytophthora infestans*, the mould that causes potato blight

botulism severe form of food poisoning caused by a neurotoxin produced by the spore forming bacterium *Clostridium botulinum*

Brill-Zinsser disease recrudescent typhus

bubas a non-venereal syphilis-like disease

bubonic plague disease caused by *Y. pestis*

bubos lymph gland swellings usually but not exclusively found in bubonic plague

bunt a disease of wheat caused by the fungus *Tilletia caries*

camelpox virus a poxvirus of camels causing severe disease with pock-like skin lesions, which is up to 25 per cent fatal

Candida albicans a yeast which is part of the normal body flora but can cause superficial infections like thrush

canine distemper virus a morbillivirus which causes distemper in dogs and some large cats

chickenpox acute infection characterized by a rash caused by *varicella zoster virus*

chitinous horny covering forming the exoskeleton of arthropods and other organisms

chlamydia sexually transmitted infection caused by *Chlamydia trachomatis* (which can also cause eye and lung infections)

chancre genital ulcer of primary syphilis caused by *Treponema pallidum*

chloroplast chlorophyll-containing structure in plant cells responsible for photosynthesis

chromosome threadlike structures of DNA and protein found in the nuclei of cells, which carry the genes

cinchona (fever) tree native of South America, the bark was used to treat malaria from early times

co-evolution linked evolution of two species usually with mutual benefit

cold sore skin lesion, usually on face around the lips, caused by *herpes simplex virus*

conacher small piece of land rented for crops or grazing in Ireland

consumption pulmonary tuberculosis

coprolite fossilized faecal material

coronavirus family of viruses containing the SARS virus and other respiratory viruses; corona, from Latin, meaning 'crown', refers to the virus's crown-like structure

Corynebacterium diphtheriae Coryneform bacteria (from Greek, meaning 'club-shaped') which causes diphtheria

cowpox pox virus which infects a wide range of domestic, zoo and wild animals; it causes lesions on cows' udders and can spread to humans

Cro-Magnon man late Palaeolithic humans named after a hill in the Dordogne, France, where their remains were found in 1868

crowd diseases the acute childhood infectious diseases caused by microbes which require a minimum number of susceptible people in close contact to maintain their chains of infection

curl a plant disease caused by the fungus *Taphrina deformans* in which the leaves curl up

cyanobacteria free-living bacteria capable of photosynthesis (previously called blue-green algae)

cytokines soluble factors produced by immune cells which regulate immune responses

cytokine storm a massive release of inflammatory cytokines in response to over-stimulation of the immune system

Darwinian evolution inherited change driven by natural selection

diatome microscopic, unicellular alga with siliceous cell wall found in plankton

diphtheria acute infectious disease caused by *Corynebacterium diphtheriae*

DNA deoxyribonucleic acid, a self-replicating molecule which carries inherited genetic information in almost all living things

DOTS directly observed therapy—short-course; a World Health Organisation programme for the treatment of TB

Duffy blood group a red blood cell surface protein that acts as a receptor for *Plasmodium vivax*

dysentery severe diarrhoea with blood and mucus which may be caused by either amoebic or bacillary infection of the large intestine

Ebola virus a filovirus (from Latin *filum* meaning thread and referring to the viruses' filamentous structure) that causes Ebola haemorrhagic fever; named after the Ebola River in Zaire near Yambuku, where the first reported outbreak occurred

ecosystem a self-sustaining community of interacting organisms

elephantiasis swollen leg(s) caused by the mosquito-transmitted filarial worm *Wuchereria bancrofti* which blocks the lymphatic drainage from the lower limbs

endemic found regularly in a particular area or population

English sweats an epidemic disease of the sixteenth century of unknown cause

engrafting (or ingrafting) inoculating with smallpox 'scabs' to induce immunity

epidemic a large-scale temporary increase in a disease in a community or region

epidemiology the study of the incidence and distribution of diseases

Epstein-Barr virus a virus that causes infectious mononucleosis (glandular fever) and is associated with a variety of human tumours (the virus is named after the scientists Anthony Epstein and Yvonne Barr, who discovered it)

erythrotherapy the red treatment used for smallpox from the twelfth to the early twentieth century

eukaryote a member of the *Eukarya* domain which includes all living things except bacteria and archaea

extremophile bacteria that exist in extreme physical conditions such as those that live under extreme pressure (barophiles), heat (thermophiles), salinity (halophiles), cold (psychrophiles)

Fertile Crescent the geographical area in modern-day Iran and Iraq between the rivers Euphrates and Tigris where agriculture was first established

filarial worm see elephantiasis

flagellum a filamentous appendage which acts as an organ of locomotion

flu (influenza) an acute infection caused by the influenza virus, an orthomyxovirus with a segmented RNA genome

flukes trematode worms with intermediate snail hosts including the causative agent of schistosomiasis

framboesia a raspberry-like lesion found in early yaws infection

gas gangrene wound infection with gas production and tissue death caused by clostridia bacteria, usually *C. perfringens*

gene the part of a chromosome, usually DNA, that codes for a specific protein

genital herpes persistent *herpes simplex virus* infection, causing recurrent genital lesions

genome the genetic material of an organism

gerbilpox a poxvirus of gerbils

gonorrhoea a sexually transmitted disease caused by the bacterium *Neisseria gonorrhoeae*

gumma the destructive inflammatory lesions of late syphilis caused by *Treponema pallidum*

HAART highly active antiretroviral therapy used to treat HIV infection

haemaglutinin a flu virus surface protein that acts as the viral receptor and induces an immune response

haemoglobin the red-coloured oxygen carrying protein found in the red blood cells of vertebrates

haemophilus influenzae a bacterium that can cause meningitis, pneumonia, septic arthritis, bronchitis and otitis media

haulm a stalk or stem

hepatitis B a hepadnavirus; a major cause of chronic liver disease and liver cancer

herpesvirus a family of viruses including *herpes simplex* and *varicella zoster* viruses

heterozygous an individual with two different copies of a particular gene

HIV human immunodeficiency virus, a retrovirus

Homo erectus Homo species dating from around 1.7 million years ago

Homo sapiens modern man dating from 150,000–200,000 years ago

hominids members of the *Homo* genus

homozygous an individual with two identical copies of a particular gene

hookworms nematode parasitic worms, *Ancylostoma duodenale* and *Necator americanus,* cause intestinal infestations common in tropical and sub-tropical regions

hyphae the branching filaments of a mould

Ice Ages prolonged periods of downturn in world temperature. The last Ice Age began to recede *c.*20,000 years ago and ended *c.*10,000 years ago; the Little Ice Age was a period of cooling in Europe lasting approximately from the thirteenth to the seventeenth century

incubation period the period between infection and the onset of symptoms

influenza see flu

inoculation the original meaning was to infection with a small dose of smallpox virus to induce immunity without severe disease, but the term is now used more widely to cover injection with infectious materials

interferons a family of cytokines some of which have anti-viral properties

Irish Apple a type of potato that stores well

louse (*phthiraptera*) wingless insect; the human parasites suck blood

Lumper a highly productive type of potato

lymph glands tissues composed of lymphocytes and other immune cells

lymphocytes cells of the immune system that are found in blood and lymph glands; B lymphocytes produce antibodies; T lymphocytes can kill virus infected cells

lysozyme a weak antibacterial substance produced by the cells of the body and present in secretions like tears

macrophage an immune cell found in tissues where it engulfs and destroys foreign and dead material; it produces cytokines that initiate an immune response

malaria a disease caused by infection with the protozoa *Plasmodium* and spread by the mosquito

meningitis infection of the membranes (meninges) surrounding the brain

mesolithic the middle phase of the Stone Age lasting from approximately 10,000 BC to the beginning of the farming era

Mesopotamia the area between the Euphrates and Tigris Rivers mostly in present-day Iraq and Iran

measles an acute infectious disease caused by measles virus, a morbillivirus; German measles (rubella) is caused by a togavirus, rubella virus

miasma Greek for 'pollution', 'bad air' or 'a noxious vapour'

mitochondria organelles found in most animal cells responsible for respiration and the generation of energy

molecular clock a measurement of the molecular differences between the genomes of two species as a way of assessing the evolutionary distance between them

molecular genetic probes labelled pieces of DNA or RNA which bind specifically to, and detect, complementary sequences

monkeypox a pox virus carried by African rodents that can infect humans

morbillivirus a genus containing the measles, canine distemper and rinderpest viruses

most recent common ancestor the most recent individual from whom a population of individuals is derived

MRSA methicillin–resistant *Staphylococcus aureus*

mumps an acute infection typified by swelling of the parotid salivary glands, caused by *mumps virus*, a paramyxovirus

Mycobacterium leprae the bacterium that causes leprosy

Mycobacterium tuberculosis the bacterium that causes tuberculosis (TB)

nagana a wasting disease of cattle caused by the trypanosome *T. brucei brucei*

natural selection survival of the fittest, leading to the propagation of their inherited characteristics

negro lethargy an early term for trypanosomiasis

neuraminidase a flu virus surface protein that induces an immune response

Palaeolithic the early Stone Age, lasting from the invention of the first tools around 35,000 BC until the end of the last Ice Age around 10,000 years ago

pandemic a disease outbreak that spreads beyond a single country

pan troglodytes troglodytes a subspecies of chimpanzee from which HIV 1 probably transferred to humans

parasite an organism living in or on another and benefiting at its expense

pathogen an organism that causes disease

Penicillium notatum the fungus that produces penicillin

Phytophthora infestans the potato blight mould

pinta a disfiguring skin infection caused by *Treponema carateum*

plague focus an area of the world where *Y. pestis* circulates among wild rodents

plasmid a circular extra-chromosomal DNA molecule which carries genetic information

plasmodium a protozoan which causes malaria including four human parasites: *P. falciparum, P. vivax, P. ovale, P. malariae,* and parasites of non-human primates such as *P. reichenowi* and *P. cynomlgi*

pneumococcus (*Streptococcus pneumoniae*) a bacterium which often harmlessly inhabits the nose and throat, but can also cause ear, sinus and lung infections as well as arthritis, peritonitis, endocarditis and meningitis

pneumonic plague fatal lung infection caused by *Y. pestis* and spread directly from person to person

pocks the skin lesions of smallpox caused by *V. major*

polio (*poliomyelitis*) a flaccid paralysis occasionally caused by poliovirus (the infection is usually asymptomatic or causes a transient meningitis)

polymorph (polymorphonuclear leucocytes) circulating immune cells with lobed nuclei and granules in their cytoplasm which contain antimicrobial substances

population bottleneck a point in time at which the population of a particular organism was very small, perhaps just one individual, from which all more recent forms arose

post-mortem examination examination after death

prokaryote all bacteria are prokaryotes which possess a simpler (prokaryotic) form of cellular organization than eukaryotes

protozoa non-photosynthetic, unicellular, microscopic organisms. Most are free-living but a few, like the malaria microbes, are parasitic

Qinghao (artemisia): ancient Chinese herbal remedy from *Artemisia annua*, the sweet wormwood, active against malaria

quarantine isolation following contact with an infectious disease sufferer, traditionally for forty days

R the case reproduction number of an epidemic, i.e. the average number of new cases derived from a single case during an epidemic

R_0 the basic reproduction number of an infectious disease, i.e. the average number of new cases derived from a single case in a susceptible population

rabies virus a *Rhabdovirus* (from the Greek meaning rod), genus *lyssavirus* (from the Greek meaning madness)

**Rattus rattus* the black rat, the intermediate host of *Y. pestis*, the cause of bubonic plague

receptor a docking molecule for attachment of a microbe or chemical to a cell

Rickettsia intracellular parasitic bacteria spread by arthropod vectors; *R. prowazekii* is the cause of typhus in humans; *R. typhi* causes murine typhus

rigor a shivering attack caused by a rising fever

rinderpest virus a paramyxovirus of the genus morbillivirus that causes rinderpest, an acute infection of cattle with a high mortality

RNA ribonucleic acid, a nucleic acid that makes up the genome of some viruses; in cells messenger RNA is transcribed from DNA and then translated into protein

rotavirus wheel-like RNA viruses (*rota* is Latin for wheel) which cause epidemics of vomiting and diarrhoea, particularly in young children

roundworm nematodes found in the intestine (such as the common roundworm, *Ascaris lumbricoides*) or tissues (such as *Trichinella spiralis*)

rust a disease of plants giving rust-coloured spots caused by fungi of the order *Uredinales*

Salmonella enterobacteria of the species *Salmonella enterica* which live in the intestine of animals and can cause outbreaks of food poisoning; *S. enterica Typhus* is the cause of typhoid fever

saprophyte a plant or microbe that lives on dead or decayed organic matter

SARS severe acute respiratory syndrome

scab a disease of potato tubers caused by *Streptomyces scabies*

scarlet fever an acute infectious disease characterized by pharyngitis and rash caused by pyrogenic exotoxin producing *Streptococcus pyogenes*

schistosome a trematode fluke which causes schistosomiasis; unusually for trematodes schistosomes are differentiated into male and female forms, the name being derived from the Greek words *schostos* and *soma*, meaning split body and denoting the groove on the male in which he holds the female

scrofula glandular tuberculosis

septicaemia severe disease caused by bacteria in the bloodstream

sex pilus a tube-like structure that grows from one bacterium and attaches to another, initiating the process of conjugation

shingles a vesicular rash confined to one dermatome caused by the herpes virus *varicella-zoster*

sickle-cell anaemia an inherited disease caused by a mutation in the haemoglobin gene which gives rise to sickle-shaped red blood cells which are destroyed rapidly leading to anaemia; carriers of the abnormal gene are protected against severe malaria

sleeping sickness see trypanosomiasis

smallpox a severe acute infectious disease characterized by skin pocks and caused by the pox virus, *Variola major*

Solanum tuberosum the potato plant

Staphylococcus a genus of bacteria named after their round shape (Greek, *kokkos*, means 'grain' or 'berry') and their cluster-forming habit (Greek, *staphyl*, meaning 'bunch of grapes'); *S. aureus* (named after the gold colour colonies it produces when grown on agar) is the major pathogen within the genus

Streptococcus pneumoniae see pneumococcus

stromatolites coral-like structures made up of colonies of interdependent bacteria, also known as microbial mats

superspreader an individual who spreads a pathogenic microbe to more than the average number of susceptible hosts

symbiosis an advantageous interaction between two different organisms living in close physical association

syphilis a chronic invasive disease caused by the spirochete *Treponema pallidum* which is generally acquired by sexual transmission or congenital infection

tapeworm a cestode which is acquired by eating undercooked contaminated meat and alternates between man and an intermediate host; *Taenia saginata*, the beef tapeworm, and *T. solium*, the pork tapeworm, are the commonest human infections

tetanus a generally fatal disease typified by muscle spasms and stiffness (also called 'lockjaw'), caused by the bacterium *Clostridium tetani*

thalassaemia a disease caused by an inherited mutation in the haemoglobin molecule that causes anaemia; carriers are protected from severe malaria

thrush a superficial infection with the yeast fungus *Candida albicans*, which may affect the mouth, gut, vagina or skin, causing white patches

toxogenic containing a toxin gene

Treponema pallidum see syphilis

Trichinella see roundworms

trypanosome a genus of protozoa that contains the causative organisms of trypanosomiasis

trypanosomiasis (sleeping sickness) a fatal parasitic disease endemic in Africa caused by *T. b. gambiense* and *T. b. rhodesiense*, both of which are spread by the tsetse fly

typhoid see salmonella

typhus a severe acute infectious disease characterized by a rash and deteriorating mental state, caused by the louse-borne Rickettsia, *R. prowazekii*

vaccination the term was originally used for immunization against smallpox (with vaccinia virus) but is now used more widely to mean immunization in general

vaccine live attenuated, killed, or subunit preparations of pathogenic microbes used to induce an immune response after immunization

varicella-zoster **virus** the herpes virus that causes chickenpox and shingles

Variola major **virus** the pox virus that causes smallpox; *V. minor* (or alastrim) is a closely related virus that causes less severe disease

Vibrio cholerae the bacterium that causes cholera

virulence the degree of pathogenicity of a microbe, as indicated by its ability to invade, damage the tissues and kill the host

West Nile fever virus a flavivirus (from the Latin, *flavus*, meaning 'yellow,' referring to the yellow fever virus which was the first of the family to be isolated) that causes West Nile Fever; West Nile fever is generally a mild disease but the virus may rarely cause encephalitis

whooping cough (*Pertussis*) an acute childhood infection caused by the bacterium *B. pertussis* (per-tussis meaning severe cough), characterized by paroxysms of coughing ending in a high-pitched whoop

Xenopsylla cheopis the rat flea

yaws a chronic skin disease endemic in certain tropical and subtropical rural populations, caused by the spirochete *T. pertenue*

yellow fever virus a flavivirus (see West Nile fever virus for derivation) found in Africa and South America, which has a reservoir in monkeys, is spread by mosquitoes and causes yellow fever in humans

Yersinia pestis primarily a flea-borne infection of rodents, *Y. pestis* causes plague in humans; *Y. pestis* evolved from *Y. pseudotuberculosis*, a gut pathogen of rats and other mammals which can also infect humans

zoonosis a natural animal pathogen, such as rabies, that on occasions infects humans

Notes and References

INTRODUCTION

1. Yu, I.T.S., Li, Y., Wong, T.W. et al., Evidence of airborne transmission of the severe acute respiratory syndrome virus. *New Engl J Med* 350: 1731–9. 2004

2. Poutanen, S.M., Low, D.E., Bonnie, H. et al., Identification of severe acute respiratory syndrome in Canada. *New Engl J Med* 348: 1995–2005. 2003

3. Reilley, B., Van Herp, M., Sermand, D., Dentico, N., SARS and Carlo Urbani. *New Engl J of Med* 348: 1951–2. 2003

4. Guan, Y., Zheng, B.J., He, Y.Q. et al., Isolation and characterization of viruses related to the SARS coronavirus from animals in Southern China. *Science* 302: 276–8. 2003

5. Yu, D., Li, H., Xu, R. et al., Prevalence of IgG antibody to SARS-associated coronavirus in animal traders—Guangdong Province, China, 2003. *Morb mort wkly* 52: 986–7. 2003

CHAPTER I: HOW IT ALL BEGAN

1. Curtis, T.P. and Sloan, W.T., Exploring microbial diversity—a vast below. *Science* 309: 1331–3. 2005

2. Suttle, C.A., Viruses in the sea. *Nature* 437: 356–61. 2005

3. Postgate, J., in *Microbes and Man*, p.13. Pelican: 1976

4. Krarhenbuhl, J.-P. and Corbett, M., Keeping the gut microflora at bay. *Science* 303: 1624–5. 2004

5. Taylor, L.H., Latham, S.M., Woolhouse, M.E., Risk factors for human disease emergence. *Phil. Trans. R. Soc. Lond.* B 356: 983–9. 2001

6. Petersen, L.R. and Hayes, E.B., Westward ho?—the spread of West Nile virus. *New Engl J Med* 351: 2257–9. 2004

7. Shild, Randy, in *And the Band Played On*, p.147. Penguin Books: 1987

CHAPTER 2: OUR MICROBIAL INHERITANCE

1. Cohen, M.N., in. *Health and the Rise of Civilisation*, p.139. Yale University Press: 1989

2. Black, F.L., Infectious diseases in primitive societies. *Science* 187: 515–8. 1975

3. Snowden, F.M., in *The Conquest of Malaria Italy, 1900–1962*, p.93. Yale University Press: 2006

4. Ross, R., On some peculiar pigmented cells found in two mosquitos fed on malarial blood. *Brit Med J* (Dec. 18): 1786–8. 1897

5. Greenwood, B. and Mutabingwa, T., Malaria in 2002. *Nature* 415: 670–2. 2002

6. Carter, R. and Mendis, K.N., Evolutionary and historical aspects of the burden of malaria. *Clin Microbiol Reviews* 15: 564–94. 2002

7. Escalantes, A.A. and Ayala, F.J., Phylogeny of the malarial genus Plasmodium, derived from rRNA gene sequences. *Proc Natl Acad Sci USA* 91: 11373–7. 1994

8. Rich, S.M., Licht, M.C., Hudson, R.R., Ayala, F.J., Malaria's Eve: evidence of a recent population bottleneck throughout the world populations of Plasmodium falciparum. *Proc Natl Acad Sci* 95: 4425–30. 1998

9. Mu, J., Duan, J., Makova, K.D. et al., Chromosome-wide SNPs reveal an ancient origin for Plasmodium falciparum. *Nature* 418: 323–6. 2002

10. See n.6 above

11. Cox, F.E.G., History of sleeping sickness (African trypanosomiasis). *Infect Dis Clin N Am* 18: 231–45. 2004

12. Welburn, S.C., Fevre, E.M., Coleman, P.G. et al. Sleeping sickness: a tale of two diseases. *Trends in Parasitology* 17: 19. 2001

13. Cohen, M.N., in *Health and the Rise of Civilisation*, p.127–8. Yale University Press: 1989

CHAPTER 3: MICROBES JUMP SPECIES

1. McNeill, W.H., in *Plagues and Peoples*, p.54

2. Diamond, J., in *Guns, Germs and Steel*, p.93–103. Vintage: 1998

3. Cohen, M.N., in *Health and the Rise of Civilisation*, p.116–22. Yale University Press: 1989

4. Cox, F.E.G., History of human parasitic diseases. *Infect Dis Clin N AM* 18: 171–88. 2004

5. Sanderson, A.T. and Tapp, E., Diseases in ancient Egypt, in *Mummies, Diseases and Ancient Cultures*, pp.38–58, eds A. Cockburn, E. Cockburn, T.A. Reyman, 2nd edn. Cambridge University Press: 1998

6. Sharp, P.M., Origins of human virus diversity. *Cell* 108: 305–12. 2002

7. Black, F.L., Measles endemicity in insular populations: critical community size and its evolutionary implication. *J Theoret Biol* 11: 207–11. 1966

8. Exod. 9: 10

9. 1 Sam. 5: 1–21

10. Brier, B., Infectious diseases in ancient Egypt. *Infect Dis Clinic N Am* 18: 17–27. 2004

11. Massa, E.R., Cerutti, N., Savoia, A.M., Malaria in ancient Egypt: paleoimmunological investigation on predynastic mummified remains. *Chungara (Arica)* 32: 7–9. 2000

12. See n.10 above

13. Mahmoud, A.A.F., Schistosomiasis (bilharziasis): from antiquity to the present. *Infect Dis Clinic N Am* 18: 207–18. 2004

14. Ibid.

15. Brant, S.V. and Loker, E.S., Can specialized pathogens colonize distantly related hosts? Schistosome evolution as a case study. *PLOS Pathogens* 1: 167–9. 2005

16. Cunha, B.A., The cause of the plague of Athens: plague, typhoid, smallpox, or measles? *Infect Dis Clinic N Am* 18: 29–43. 2004

17. Ibid.
18. Zinsser, H., in *Rats, Lice and History*, p.121. Blue Ribbon Books, Inc.: 1934
19. Cunha, B.A., The death of Alexander the Great: malaria or typhoid fever? *Infect Dis Clinic N Am* 18: 53–63. 2004
20. Fears, J.F., The plague under Marcus Aurelius and the decline and fall of the Roman Empire. *Infect Dis Clinic N Am* 18: 65–77. 2004
21. Ibid.
22. Zinsser, H., in *Rats, Lice and History*, pp.146–7. Blue Ribbon Books, Inc.: 1934

CHAPTER 4: CRAWDS, FILTH AND POVERTY

1. Mc Neill, W.H., in *Plagues and Peoples*, p.80. Anchor Books: 1976
2. Scott, S., and Duncan, C., in *Return of the Black Death*, pp.14–15. Wiley: 2004
3. Benedictow, O.J., in *The Black Death 1346–1353*, p.142. BCA: 2004
4. Scott, S., and Duncan, C., in *Return of the Black Death*, p.49. Wiley: 2004
5. Ziegler, P., in *The Black Death*, pp.116–17. Penguin Books: 1969
6. Ziegler, P., in *The Black Death*, p.67. Penguin Books: 1969
7. Pepys, S., in *The Diary of Samuel Pepys: A Selection*, p.1665, ed R. Latham. Penguin Books: 1985
8. Kitasato, S., The bacillus of bubonic plague. *The Lancet* (25 August): 428–30. 1894
9. Yersin, A., La peste bubonique a Hong Kong. *Ann Inst Pasteur* 8: 662–7. 1894
10. Hinnebusch, B.J., The evolution of flea-borne transmission in *Yersinia pestis*. *Curr Issues Mol Biol* 7: 197–212. 2005
11. Duncan, C.J. and Scott, S., What caused the Black Death? *Postgrad Med J* 81: 315–20. 2005
12. Orent, W., in *Plague*, p. 123. Free Press: 2004
13. Scott, S. and Duncan, C., in *Return of the Black Death*, p. 195. Wiley: 2004
14. Benedictow, O.J., in *The Black Death 1346–53*, p.22. BCA: 2004
15. Raoult, D., Aboudharam, G., Crubezy, E. et al., Molecular identification by 'suicide PCR' of Yersinia pestis as the agent of medieval black death. *Proc Natl Acad Sci USA* 97: 12800–3. 2000

16. Gilbert, M.T.P., Cuccui, J., White, W. et al., Absence of *Yersinia pestis*-specific DNA in human teeth from five European excavations of putative plague victims. *Microbiology* 150: 341–54. 2004

17. Scott, S. and Duncan, C., in *Return of the Black Death*, p.225. Wiley: 2004

18. Bradbury, J., Ancient footsteps in our genes: evolution and human disease. *The Lancet* 363: 952–3. 2004

19. Benedictow, O.J., in *The Black Death 1346–53*, p.16. BCA: 2004

20. Ibid., pp.387–94

21. Gubser, C., Hue, S., Kellam, P., Smith, G.L., Poxvirus genomes: a phylogenetic analysis. *J Gen Virol* 85: 105–17. 2004

22. Hopkins, D.R., in *Princes and Peasants*, pp.14–15. University of Chicago Press: 1983

23. See Ch. 3, n.5.

24. Hopkins, D.R., in *Princes and Peasants*, p.24. University of Chicago Press: 1983

CHAPTER 5: MICROBES GO GLOBAL

1. McNeill, W.H., in *Plagues and Peoples*, p.214. Anchor Books: 1976

2. Crosby, A.W., in *The Columbian Exchange*, p.36. Greenwood Press: 1972

3. Ibid., p.56

4. Diamond, J., in *Guns, Germs and Steel*, pp.70–1. Vintage: 1998

5. Crosby, A.W., in *The Colombian Exchange*, p.51. Greenwood Press: 1972

6. Dickerson, J.L., in *Yellow Fever*, pp.13–32. Prometheus Books: 2006

7. Ibid., pp.141–86

8. Hyden, D., in *Pox: Genius, Madness and the Mysteries of Syphilis*, p.13. Basic Books: 2003

9. Pusy, W.A., in *The History and Epidemiology of Syphilis*, p.8. C.C. Thomas: 1933

10. Hyden, D., in *Pox: Genius, Madness and the Mysteries of Syphilis*, p.22. Basic Books: 2003

11. Tramont, E.C., The impact of syphilis on humankind. *Infect Dis Clin N Am* 18:101–10. 2004

12. Hyden, D., in *Pox: Genius, Madness and the Mysteries of Syphilis*, p.12. Basic Books: 2003

13. Von Hunnius, T.E., Roberts, C.A., Boylston, A., Saunders, S.R., Historical identification of syphilis in Pre-Columbian England. *Am J Phys Anthropol* 129: 559–66. 2006

14. Rothschild, B.M., History of syphilis. *CID* 40: 1454–63. 2005

15. Fraser, C.M., Norris, S.J., Weinstock, G.M., White, O. et al., Complete genome sequence of Treponema pallidum, the syphilis spirochete. *Science* 281: 375–88. 1998

16. See n. 11 of this chapter.

17. Pollitzer, R., in *Cholera*, p.18. WHO monograph: 1959

18. De, S.N., in *Cholera, its Pathology and Pathogenesis*, pp.10–11. Oliver and Boyd: 1961

19. Faruque, S.M., Bin Naser, I., Islam, M.J., et al., Seasonal epidemics of cholera inversely correlate with the prevalence of environmental cholera phages. *Proc Natl Acad USA* 102: 1702–07. 2005

20. Siddique, A.F., Salam, A., Islam, M.S., et al., Why treatment centres failed to prevent cholera deaths among Rwandan refugees in Goma, Zaire. *The Lancet* 345: 359–61. 1995

21. Markel, H., in *When Germs Travel*, p.201. Pantheon Books: 2004

CHAPTER 6: FAMINE AND DEVASTATION

1. Zuckerman, L., in *The Potato: from the Andes in the sixteenth century to fish and chips, the story of how a vegetable changed history*, p.19. Macmillan: 1999

2. Ibid., p 31

3. Large, E.C., in *The Advance of the Fungi*, p.24. Jonathan Cape: 1940

4. Ibid., p23

5. Zuckerman, L., in *The Potato: from the Andes in the sixteenth century to fish and chips, the story of how a vegetable changed history*, p.187. Macmillan: 1999

6. Large, E.C., in *The Advance of the Fungi*, p.13. Jonathan Cape: 1940

7. Zuckerman, L., in *The Potato: from the Andes in the sixteenth century to fish and chips, the story of how a vegetable changed history*, p.186. Macmillan: 1999

8. Ibid., p.189

9. Ibid., p.190

10. Large, E.C., in *The Advance of the Fungi*, p.34. Jonathan Cape: 1940

11. Zuckerman, L., in *The Potato: from the Andes in the sixteenth century to fish and chips, the story of how a vegetable changed history*, p.191. Macmillan: 1999

12. Large, E.C., in *The Advance of the Fungi*, p.38. Jonathan Cape: 1940

13. Zuckerman, L., in *The Potato: from the Andes in the sixteenth century to fish and chips, the story of how a vegetable changed history*, p.188. Macmillan: 1999

14. Ibid., p.194

15. Ibid., p.198

16. Large, E.C., in *The Advance of the Fungi*, p.38. Jonathan Cape: 1940

17. Ibid., p.20

18. Berkley, M.J., Observations, botanical and physiological, on the potato murrain. *J Hortic Soc Lond* 1: 9–34.1846

19. Large, E.C., in *The Advance of the Fungi*, p.27. Jonathan Cape: 1940

20. Ibid., p.40

21. Ibid., p.20

22. McLeod, M.P., Qin, X., Karpathy, S.E. et al., Complete genome sequence of *Rickettsia typhi* and comparison with sequences of other Rickettsiae. *J Bact* 186: 5842. 2004

23. Zinsser, H., in *Rats, Lice and History*, pp.161–4. Blue Ribbon Books, Inc: 1934

24. McDonald, P., in *Oxford Dictionary of Medical Quotations*. Oxford University Press: 2004

25. McNeill, W.H., in Plagues and Peoples, p.278 Anchor Books: 1976

26. Daniels, T.M., The impact of tuberculosis on civilization. *Infect Dis Clin N Am* 18: 157–65. 2004

27. Brosch, R., Gordon, S.V., Marmiesse, M. et al., A new evolutionary scenario for the *Mycobacterium tuberculosis* complex. *Proc Nalt Acad Sci USA* 99: 3684–9. 2002.

28. See n. 26 of this chapter.

CHAPTER 7: DEADLY COMPANIONS REVEALED

1. Duran-Reynals, M.L., in *The Fever Tree: the pageant of quinine*, pp.34–5. W.H. Allen, London: 1947

2. <http://www.bbc.co.uk/history>

3. Bassler, B.L. and Losick, R., Bacterially Speaking. *Cell* 125: 237–46. 2006

4. Fenner, F., Henderson, D.A., Arita, I., Jezek, Z., Ladnyi, I.D., in *Smallpox and its Eradication*, pp.252–3. World Health Organisation, Geneva: 1988

5. Hopkins, D.R., in *Princes and Peasants*, p.46. University of Chicago Press: 1983

6. Ibid., pp.47–8

7. Ibid., p.47

8. Halsband, R. New light on Lady Mary Wortley Montagu's Contribution to Inoculation. *J Hist Med and Allied Sciences* 8: 309–405. 1953

9. Hopkins, D.R., in *Princes and Peasants*, p.50. University of Chicago Press: 1983

10. Ibid., p.79

11. Ibid., p.85

12. Ibid., p.95

13. Ibid., p.80

14. Fenner, F., Henderson, D.A., Arita, I., Jezek, Z., Ladnyi, I.D., in *Smallpox and its Eradication*, pp.264–5. World Health Organisation: 1988

15. Alibek, K., in *Biohazard* p.261. Hutchinson: 1999

16. Fleming, A., On the antibacterial action of cultures of a penicillium, with special reference to their use in isolation of *B influenzae*. *Brit J Exper Path* 10: 226–36. 1929

17. Macfarlane, G., in *Alexander Fleming: the man and the myth*, p.130. The Hogarth Press: 1984

18. Ibid., p.164

19. Chain, E., Florey, H.W., Gardner, A.D. et al., Penicillin as a chemotheraoeutic agent. *The Lancet* ii: 226–8. 1940

20. Macfarlane, G., in *Alexander Fleming: the man and the myth*, p.178. The Hogarth Press: 1984

CHAPTER 8: THE FIGHT BACK

1. Coale, A.J., The history of the human population, in *Biological Anthropology* (readings from *Scientific America*), ed. Katz, S., pp.659–70. W.H. Freeman & Co, San Francisco: 1075

2. Heeney, J.L., Dalgleish, A.G., Weiss, R.A., Origins of HIV and the evolution of resistance to AIDS. *Science* 313: 462–6. 2006

3. Avasthi, A., Bush-meat trade breeds new HIV. *New Scientist* (7 August): 8 2004

4. Reed, K.D. J. W., Melski, MB., Graham et al., The detection of Monkeypox in humans in the Western hemisphere. *New Engl J Med* 350: 342–50. 2004

5. See n. 3 of this chapter

6. McMichael, T., in *Human Frontiers, Environments and Disease: past patterns, uncertain futures,* p.95. Cambridge University Press: 2001

7. Cliff, A. and Haggett, P., Time, travel and infection.*Brit med Bulletin* 69: 87–99. 2004

8. Bradley, D.J., The scope of travel medicine, in *Travel Medicine: proceedings of the first conference on international travel medicine,* pp.1–9. Springer Verlag: 1989

9. Coghlan, A., Jet-setting mozzie blamed for malaria case. *New Scientist* (31 August): 9. 2002

10. Newton, G. (ed.), In *Antibiotic Resistance an Unwinnable War?,* p.2. Wellcome Focus: 2005

11. Ibid., p.26

12. Cohen, J., Experts question danger of 'AIDS superbug'. *Science* 307: 1185. 2005

13. Gandy, M. and Zumla, A. (eds), *The Return of the White Plague: global poverty and the 'new' tuberculosis,* p.129. Verso: 2003

14. Klausner, R. and Alonso, P., An attack on all fronts, *Nature Outlook* supplement, *Malaria the Long Road to a Healthy Africa,* pp.930–1. 2004

15. Attaran, A., Where did it all go wrong?, *Nature Outlook* supplement, *Malaria the Long Road to a Healthy Africa,* pp.932–3. 2004

16. Garcia-Sastre, A. and Whitley, R.J., Lessons learned from reconstructing the 1918 influenza pandemic. *JID* 194 (Suppl.2): ps127–s132. 2006

17. Tumpey, T.M., Basler, C.F., Aguilar, C.F. et al., Characterisation of the reconstructed 1918 Spanish influenza pandemic virus. *Science* 310: 77–80. 2005

18. Seo, S.H., Hoffmann, E., Webster, R.G., Lethal H5N1 influenza viruses escape host antiviral cytokine responses. *Nature Medicine* 8: 950–4. 2002

19. Fauci, A.S., Emerging and re-emerging infectious diseases: influenza as a prototype of the host-pathogen balancing act. *Cell* 124: 665–70. 2006

20. Mackenzie, D., Animal apocalypse. *New Scientist* (13 May): 39–43. 2006

CONCLUSION: LIVING TOGETHER

1. *Treating Infectious Diseases in a Microbial World*, Report of two workshops on novel antimicrobial therapeutics p.1. National Academies Press: 2006

Further Reading

INTRODUCTION

Abraham, Thomas, *Twenty-first century plague—The Story of SARS*. Johns Hopkins Press: 2004

Skowronski, D.M., Astell, C., Brunham, R.C. et al., Severe acute respiratory syndrome (SARS): a year review. *Annu.Rev.Med.* 56: 357–81. 2005

CHAPTER 1: HOW IT ALL BEGAN

Cockell, C., *Impossible Extinctions*. Cambridge University Press: 2003

Dronamraju, K.R., *Infectious Disease and Host-Pathogen Evolution*. Cambridge University Press: 2004

Posgate, J., *Microbes and Man*. Pelican: 1976

CHAPTER 2: OUR MICROBIAL INHERITANCE

Carter, R. and Mendis, K.N., Evolutionary and Historical Aspects of the Burden of Malaria. *Clinical Microbiology Reviews* 15: 564–94. 2002

Cohen, M.N., *Health and the Rise of Civilisation*. Yale University Press: 1989

T-W Fiennes, R.N., *Zoonoses and the Origins and Ecology of Human Disease*. Academic Press: 1978

Foster, W.D., *A History of Parasitology*. E.&S. Livingstone Ltd: 1965

McNeil, W.H., *Plagues and Peoples*. Anchor Books: 1976

Maudlin, I., African trypanosomiasis. *Annals of Tropical Medicine and Parasitology* 100: 679–701. 2006

CHAPTER 3: MICROBES JUMP SPECIES

Diamond, J., *Guns, Germs and Steel*. Vintage: 1998

T-W-Fiennes, R.N., *Zoonoses and the Origins and Ecology of Human Disease*. Academic Press: 1978

Gryseels, B., Polman, K., Clerinx, J., Kestens, L., Human schistosomiasis. *The Lancet* 368: 1106–17. 2006

McNeill, W.H., *Plagues and Peoples*. Anchor Books: 1976

CHAPTER 4: CROWDS, FILTH AND POVERTY

Benedictow, O.J., *The Black Death 1346–53*. BCA: 2004

Hopkins, D.R., *Princes and Peasants*. University of Chicago Press: 1983

McNeill, W., *Plagues and Peoples*. Anchor Books: 1976

Marriott, E., *The Plague Race*. Picador: 2002

Orent, W., *Plague*. Free Press: 2004

Robinson, B., *The Seven Blunders of the Peaks*. Scarthin Books: 1994

Scott, S. and Duncan, C., *Return of the Black Death*. Wiley: 2004

CHAPTER 5: MICROBES GO GLOBAL

Bryan, C.S., Moss, S.W., Kahn, R.J., Yellow fever in the Americas. *Infect Dis Clin N Am* 18: 275–92. 2004

Crosby, A.W., *The Colombian Exchange*. Greenwood Press: 1972

Hyden, D., *Pox: Genius, Madness and the Mysteries of Syphilis*. Basic Books: 2003

Pusy, W.A., *The History and Epidemiology of Syphilis*. C.C. Thomas: 1933

Sack, D.A., Sack, R.B., Nair, G.B., Siddique, A.K., Cholera. *The Lancet* 363: 223–33. 2004

Vinten-Johansen, P., Brody, H., Paneth, N., Rachman, S., Rip, M., *Cholera, Chloroform, and the Science of Medicine*. Oxford University Press: 2003

CHAPTER 6: FAMINE AND DEVASTATION

Daniels, T.M., The impact of tuberculosis on civilization. *Infect Dis Clin N Am* 18: 157–65. 2004

Gandy, M. and Zumla, A. (eds), *The Return of the White Plague: global poverty and the 'new' tuberculosis.* Verso: 2003

Large, E.C., *The Advance of the Fungi.* Jonathan Cape: 1940

Leavitt, J.W., *Typhoid Mary: captive to the public's health.* Beacon Press: 1997

Raoult, D., Woodward, T., Dumler, J.S., The history of epidemic typhus. *Infect Dis Clin N Am* 18: 127–40. 2004

Zuckerman, L., *The Potato: from the Andes in the sixteenth century to fish and chips, the story of how a vegetable changed history.* Macmillan: 1999

CHAPTER 7: DEADLY COMPANIONS REVEALED

Alibek, K., in *Biohazard.* Hutchinson: 1999

Hopkins, D.R., in *Princes and Peasants.* University of Chicago Press: 1983

Macfarlane, G., in *Alexander Fleming: the man and the myth.* The Hogarth Press: 1984

CHAPTER 8: THE FIGHT BACK

Gandy, M. and Zumla, A. (eds), *The Return of the White Plague: global poverty and the 'new' tuberculosis.* Verso: 2003

Emerging infectious diseases. *Nature Medicine* 10 (supplement). 2004

The Lancet 367: 875–58. 2006

McMichael, T., in *Human Frontiers, Environments and Disease: past patterns, uncertain futures.* Cambridge University Press: 2001

Nature Outlook. supplement, *Malaria the Long Road to a Healthy Africa.* 2004

Index

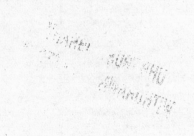